Insects: Functional Morphology, Biomechanics and Biomimetics

Insects: Functional Morphology, Biomechanics and Biomimetics

Editors

Hamed Rajabi
Jianing Wu
Stanislav N. Gorb

MDPI • Basel • Beijing • Wuhan • Barcelona • Belgrade • Manchester • Tokyo • Cluj • Tianjin

Editors
Hamed Rajabi
School of Engineering
London South Bank University
London
United Kingdom

Jianing Wu
School of Aeronautics
and Astronautics
Sun Yat-Sen University
Shenzhen
China

Stanislav N. Gorb
Functional Morphology
and Biomechanical
Kiel University
Kiel
Germany

Editorial Office
MDPI
St. Alban-Anlage 66
4052 Basel, Switzerland

This is a reprint of articles from the Special Issue published online in the open access journal *Insects* (ISSN 2075-4450) (available at: www.mdpi.com/journal/insects/special_issues/morphology_biomechanics).

For citation purposes, cite each article independently as indicated on the article page online and as indicated below:

LastName, A.A.; LastName, B.B.; LastName, C.C. Article Title. *Journal Name* **Year**, *Volume Number*, Page Range.

ISBN 978-3-0365-2893-9 (Hbk)
ISBN 978-3-0365-2892-2 (PDF)

Contents

About the Editors

Hamed Rajabi

Dr Hamed Rajabi is a Lecturer in the School of Engineering at London South Bank University (LSBU). He received his first PhD in Mechanical Engineering followed by a second one in Biology. Collaborations with researchers from different backgrounds have enabled him to employ methods and techniques from different fields into his research and, thereby, answer questions that can be addressed only using multidisciplinary approaches.

Hamed is passionate about biological systems and their 'technological' complexities. He and his team aim to unravel these complexities, learn from them to develop nature-inspired concepts, and elaborate them into a technology readiness level that can be converted into marketable products, especially in the areas of structural reinforcement, lightweight construction, healthcare, and robotics.

Hamed has contributed over 55 articles in peer-reviewed journals (many of which are ranked among the highest scored articles in all research areas), one preprint, two editorials, six articles in languages other than English (including Persian, German and Japanese), and three patents. His research has been featured by many science news magazines, such as Phys.org, New Atlas, 3D Printing Industry, TechXplore, Advanced Science News, etc. He has won multiple national and international awards, including KiNSIS Best PhD Dissertation Award, and KLS Postdoctoral Award for excellent research quality. He has supervised over 40 postgraduate students. Hamed is currently providing service as a member of the Editorial Board of Frontiers in Mechanical Engineering and Journal of Bionic Engineering, and as a Standing Member of the Youth Committee of the International Society of Bionic Engineering.

Jianing Wu

Dr Jianing Wu is an Associate Professor in the School of Aeronautics and Astronautics at Sun Yat-Sen University, China. He received his PhD in Mechanical Engineering from Tsinghua University China, in 2015. He joined Georgia Institute of Technology, USA, as a postdoctoral research fellow from 2015 to 2018, worked on biolocomotion and bio-inspired systems. Later, he joined Sun Yat-Sen University as an associate professor.

Dr Wu's research interests include animal behavior, biomechanics and bio-inspired robotics. He has published 50 papers (40 of them as the first/corresponding author), including Acta Biomaterialia, Royal Society Interface, Journal of Experimental Biology and Bioinspiration & Biomimetics. These works have been cited more than 800 times at Google Scholar. The outcomes of his works have been highlighted in many world-known science media, such as Science, New York Times, New Scientists, Science Daily, Physics Org, etc.

Stanislav N. Gorb

Stanislav Gorb is Professor and Director at the Zoological Institute of the Kiel University, Germany. He received his PhD degree in zoology and entomology at the Schmalhausen Institute of Zoology of the Ukrainian Academy of Sciences in Kiev (Ukraine). Gorb was a postdoctoral researcher at the University of Vienna (Austria), a research assistant at University of Jena, a group leader at the Max Planck Institutes for Developmental Biology in Tübingen and for Metals Research in Stuttgart (Germany). Gorb's research focuses on morphology, structure, biomechanics, physiology, and evolution of surface-related functional systems in animals and plants, as well as the development of biologically inspired technological surfaces and systems. He received the Schlossmann Award in Biology and Materials Science in 1995, International Forum Design Gold Award in 2011 and Materialica "Best of" Award in 2011. In 1998, he was the BioFuture Competition winner for his works on biological attachment devices as possible sources for biomimetics. In 2018, he received Karl-Ritter-von-Frisch Medal of German Zoological Society. Gorb is Corresponding member of Academy of the Science and Literature Mainz, Germany (since 2010) and Member of the National Academy of Sciences Leopoldina, Germany (since 2011). Gorb has authored several books, more than 500 papers in peer-reviewed journals, and five patents.

Editorial

Insects: Functional Morphology, Biomechanics and Biomimetics

Hamed Rajabi [1],* , Jianing Wu [2],* and Stanislav Gorb [3],*

[1] Division of Mechanical Engineering and Design, School of Engineering, London South Bank University, London SE1 0AA, UK
[2] School of Aeronautics and Astronautics, Sun Yat-sen University, Guangzhou 510006, China
[3] Functional Morphology and Biomechanics, Zoological Institute, Kiel University, D-24118 Kiel, Germany
* Correspondence: rajabijh@lsbu.ac.uk (H.R.); wujn27@mail.sysu.edu.cn (J.W.); sgorb@zoologie.uni-kiel.de (S.G.)

Citation: Rajabi, H.; Wu, J.; Gorb, S. Insects: Functional Morphology, Biomechanics and Biomimetics. *Insects* **2021**, *12*, 1108. https://doi.org/10.3390/insects12121108

Received: 19 November 2021
Accepted: 9 December 2021
Published: 12 December 2021

Publisher's Note: MDPI stays neutral with regard to jurisdictional claims in published maps and institutional affiliations.

1. Introduction to the Special Issue

Insects are the most diverse animal taxon, both in terms of the number of species and the number of individuals. There are roughly one million described insect species, and their real number is estimated to be five to ten times this figure [1]. The total number of individual insects, on the other hand, is estimated to be as high as one million trillion [2]. This is why insects are regarded as the most successful groups of animals on Earth.

It is not the first time, nor will it be the last, that insects have become the core theme for a collection of experimental studies. However, what makes the current collection unique is the focus on understanding the complexities of insect structures, functions and potential applications that they offer. Although the functional morphology of insects remains a basic science, insect structures offer a variety of existing and potential applications in biomedical, structural, mechanical and aerospace engineering. In this Special Issue "Insects: Functional Morphology, Biomechanics and Biomimetics", we aimed to include studies that cover fundamental research and discuss practical applications, as much as possible.

In this Special Issue, our readers will read about insect structures, including wings, legs, feeding apparatus, sensory organs and ovipositors. The reader will find answers to questions such as: How do insects fly? What are the design strategies that enable insect wings to reach automatic shape control? How does the specific segmented design determine the oscillatory response of insect antennae? What are the adaptations of insect mouthparts to their feeding habits? How does a thin film of adhesive secretion contribute to attachment of insect eggs? What are the physiological and environmental factors that determine the jump performance of locusts? How can these inspire engineering innovations, such as wings for micro air vehicles, enhanced sensory systems or multifunctional microfluidic transporters?

A number of well-known scholars in the field kindly accepted our invitation and contributed to this collection. We would like to thank the authors for their invaluable contributions to this Special Issue. We appreciate their dedication, support and commitment. We would also like to thank the support of *Insects*, MDPI and its staff, who made this Special Issue possible. We very much thank the reviewers who assessed the submitted manuscripts and played a key role in improving the quality of this Special Issue. We hope that you enjoy reading this Special Issue.

2. Dedication

This Special Issue is dedicated to Professor Leonid I. Fransevich, a corresponding member of the Ukrainian National Academy of Sciences and Professor Emeritus at Schmalhausen Institute of Zoology, Kiev, Ukraine, for his work in insect functional morphology, physiology and biomechanics, and on the occasion of his 85th birthday.

Leonid Frantsevich was born in 1935 in Kiev, Ukraine. He graduated from the Faculty of Biology of the Shevchenko Kiev State University with a diploma in Biology–Zoology.

He defended his PhD thesis "Fauna of Lepidoptera of the Middle Dnieper Valley" (1963) and his doctoral (habilitation) dissertation "Visual analysis of space in insects" (1981), both in entomology. For over 40 years, he has been working at the Schmalhausen Institute of Zoology of the Ukrainian National Academy of Sciences, where he currently holds the position of leading researcher.

Professor Fransevich made a number of important discoveries in the studies of insects. In particular, he discovered the ability of animals to recognize random two-dimensional images by their texture. He showed astro-orientation in Coleoptera during homing and identified structural elements of the olfactory center in the insect brain (glomeruli of the deutocerebrum) by morphological characteristics. He discovered, described and experimentally studied a type of proprioceptor (arcular organ) in Coleoptera. He demonstrated the spatial stability of visual orientation behind local and astro-landmarks in insects during homing on inclined surfaces. He proposed an orientation model using polarized sky light based on a standard polarization sensitivity direction map embedded in the retinal structure and demonstrated the spatial stability of topological signs of visual key stimuli in insects.

Leonid Frantsevich was the first to use the skeletal model of the kinematic system of walking insects for the purpose of describing and analyzing movements, solving the inverse kinematics problem for reconstructing the joint angles, which are not directly observed. Using inverse kinematics methods, he studied the kinematics of locomotor maneuvers in walking insects: turns on a plane, overturns, walking on thin rods, turning at the end of a thin rod, as well as the kinematics of opening–closing elytra in Coleoptera. He contributed to the research on the kinematics and mechanism of deployment of the arolium (a sticky pad at the insect pretarsus) and the role of pre-stressed structures in this process. Having studied the mechanics of the composed middle coxa in dipterans, Leonid Frantsevich showed that this structure is a marker of the body segment to which the leg is attached and discovered manifestations of homeosis (the appearance of a structure in another body segment) in certain dipteran taxa.

From autumn 1986 to 1988, Leonid Frantsevich headed the work of Kiev zoologists in the Chornobyl NPP site and the Exclusion Zone. At that time, his research developed in two directions: radioecology and general ecology. He calculated the volume of the removal of the radionuclides from the Exclusion Zone by migratory birds and then proposed an integral estimate of the offset as a product of three quantities. The resulting estimate turned out to be insignificant in comparison with the total removal of radionuclides outside the zone and did not require specific countermeasures.

In 1989–1994, Leonid Frantsevich and his colleagues carried out a wide bioindication of 90Sr pollution of water bodies and land on the basis of the beta radioactivity of mollusk shells. Maps of 90Sr pollution of the Kiev region and rivers of the Dnieper basin were compiled. The experience of data generalization for multi-species collections was used to reconstruct the radioactive pollution of various species of wild animals based on the study of a few representative species, which are accepted as comparison standards. This standardization made it possible to depict the radionuclide contamination of wild animals on a map (2000). Based on the methods of processing multicomponent collections, Leonid Frantsevich created the first model for optimizing the permissible levels of radionuclides in food (1997).

Leonid Frantsevich was the first to draw attention to the fact that, in most of the exclusion and resettlement zones (over 98% of the total area—about 3000 km^2), the course of events in biocenoses was determined not by the harmful effect of radiation, but rather by the removal of anthropogenic pressure on wildlife after the evacuation of the population, eliminating large-scale engineering interventions. Research and accounting of general ecological patterns were needed for the management of the alienated territories. He proposed the concept of a mosaic reserve of the Exclusion Zone with the allocation of scientifically or nature-protected lands. The principle of the mosaic reserve was approved by the Scientific and Technical Council under the Administration of the Exclusion Zone.

Leonid Frantsevich has been very successful and productive (over 150 original publications and related books, the most significant of which are "Visual analysis of space in insects" (1980), "Spatial orientation of animals" (1986) and "Animals in the radioactive zone" (1991). During his career, he has received several awards for his many contributions to science, including the State Prize of the USSR (1987), the State Prize of the Ukraine (2004) and being elected as a corresponding member of the Ukrainian National Academy of Sciences (1990), to name a few.

We organized this Special Issue in honor of Professor Frantsevich's distinguished scientific career over the past 60 years. This Special Issue consists of original research articles and review articles related to the functional morphology and biomechanics of insects, his favorite topic.

Funding: This research received no external funding.

Conflicts of Interest: The authors declare no conflict of interest.

References

1. Gaston, K.J. The magnitude of global insect species richness. *Conserv. Biol.* **1991**, *5*, 283–296. [CrossRef]
2. Wilson, E.O. Hotspots-Preserving pieces of fragile biosphere. *Natl. Geogr.* **2002**, *1*, 318.

Article

The Geometry and Mechanics of Insect Wing Deformations in Flight: A Modelling Approach

Robin Wootton

Department of Biosciences, University of Exeter, Address for correspondence 61 Thornton Hill, Exeter EX4 4NR, UK; r.j.wootton@exeter.ac.uk

Received: 16 June 2020; Accepted: 13 July 2020; Published: 17 July 2020

Abstract: The nature, occurrence, morphological basis and functions of insect wing deformation in flight are reviewed. The importance of relief in supporting the wing is stressed, and three types are recognized, namely corrugation, an M-shaped section and camber, all of which need to be overcome if wings are to bend usefully in the morphological upstroke. How this is achieved, and how bending, torsion and change in profile are mechanically interrelated, are explored by means of simple physical models which reflect situations that are visible in high speed photographs and films. The shapes of lines of transverse flexion are shown to reflect the timing and roles of bending, and their orientation is shown to determine the extent of the torsional component of the deformation process. Some configurations prove to allow two stable conditions, others to be monostable. The possibility of active remote control of wing rigidity by the thoracic musculature is considered, but the extent of this remains uncertain.

Keywords: insects; wings; deformation; flight; bending; torsion; camber; control; physical models

1. Introduction

This paper has a dual function: to review the occurrence of flight-related deformations in the morphological upstroke of insect wings and to investigate the geometric principles underlying the interaction of bending, torsion and camber change, by means of simple physical models.

Orthodox, flight-adapted insect wings are smart structures: they are flexible aerofoils whose three-dimensional shape from instant to instant in flight is largely determined by their elastic response to the aerodynamic and inertial forces they are receiving. While the profile of the wing base can normally be altered and controlled by the direct flight muscles of the thorax, the absence of muscles within the wing requires that three-dimensional shape control beyond the base is to a great extent automatic — encoded in the wing's detailed structure. Four decades ago, I discussed the nature and function of the deformations they undergo, and identified a range of morphological adaptations to facilitate and to limit them [1]. The extensive research carried out since then has expanded and broadly confirmed these early conclusions and predictions [2–26] (in particular, see [19,21] for summaries of the extensive Russian literature), and major advances in insect aerodynamics have greatly helped to interpret their significance, e.g., [27–35].

Our knowledge of wing kinematics and deformations has come from high speed still and cine photography and video-recording. These sources show, unsurprisingly, that in the insects studied, the wings' cyclic deformations are not rigidly determined: they vary in extent, even within a given flight sequence. To take just one example, high speed photographs of *Panorpa communis* (Mecoptera) in the upstroke published by Brackenbury [16] show virtually no bending in the wings, and Brodsky and Ivanov [4], filming tethered individuals, found little wing flexion, but a short high speed movie sequence of *Panorpa germanica* shortly after take-off shows extensive upstroke bending of the forewings and particularly the hindwings increasing from stroke to stroke [14] (Figure 1). These are different species, but their wings are structurally identical, and one would expect similar behaviour in both. These variations between strokes may be passive: wing shape must certainly be influenced by variations in angular velocity in the translation part of the stroke and in angular acceleration around stroke reversal. However, there is a possibility that, in some insects at least, a degree of control of bending, passive torsion and section may be exerted remotely by muscles at the wing base, and it is interesting to explore how such control might be achieved. Furthermore, wings, as resonant structures, need to deform appropriately at their actual flapping frequencies, and it is entirely possible that they may be tunable by active control of wing rigidity.

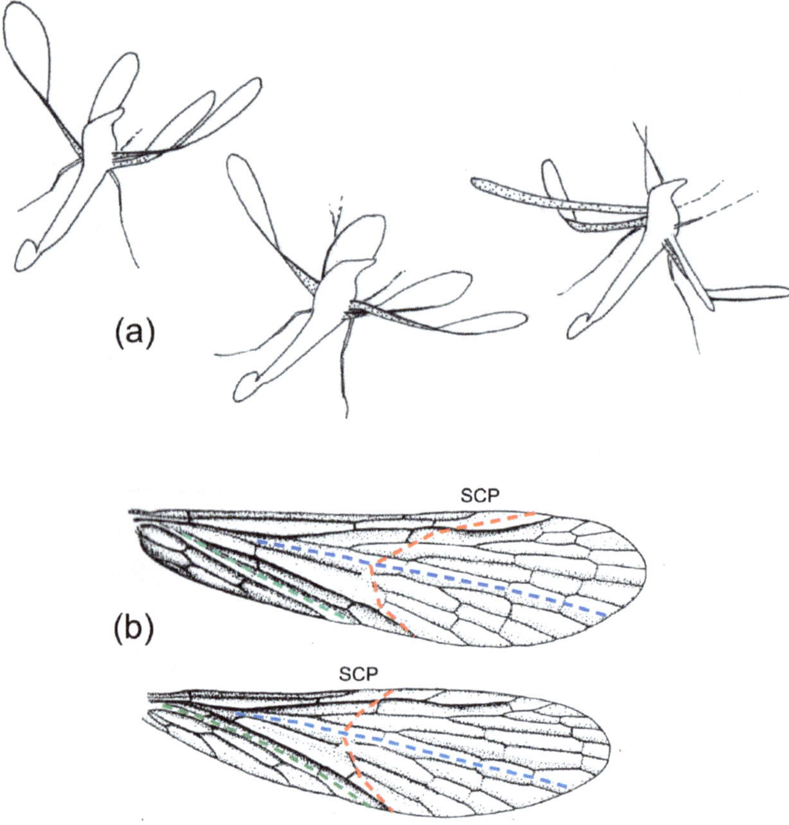

Figure 1. (**a**) Tracings of three frames from the same upstroke of *Panorpa germanica* from a high-speed film by A.R. Ennos. Note the very different bending modes of forewings and hindwings, reflecting the different lengths of the subcosta, SCP, and that flexion and torsion persist throughout the half-stroke. (**b**) Fore and hind wings of *Panorpa germanica*. Here, and in subsequent wing illustrations, the median flexion line is shown in blue, transverse flexion lines in red and the claval flexion line in green.

In the last two decades, particularly stimulated by the biomimetic possibilities in the development of micro air vehicles, there has been a great increase in interest in the structure, properties and functioning of the wings of certain groups: hawkmoths [25,26,29,36], locusts [23,37–39], hoverflies [40–42] and, above all, Odonata [43–50]; see [46–48] for reviews of the extensive literature, in which modelling has played an increasingly important role. Models have long been valuable in understanding wing functioning, and Wootton et al. [24] identified a logical sequence from conceptual though physical and analytical models to increasingly sophisticated computational simulations of individual species.

Each stage in this sequence has both advantages and limitations. Computational models now rightly dominate the literature, but they are vulnerable to incorrect initial assumptions, and, historically, some of the most useful information has come from simple physical models, based on direct observation of insects in flight and simple manipulation of wings. These are easy and quick to construct and have allowed the swift investigation and testing of a range of observed phenomena in a broad range of insects, in some cases giving direction to analytical and computational modelling of complete wings or wing components [3,12,23,24,38,47,49–55].

In 1999, I further discussed wing design, deformation and control in the wider context of invertebrate paraxial locomotory appendages, and illustrated how the principles underlying the in-flight deformation of many wings can be learned as a first approximation by modelling them as simple shells; see [55] for a wider range of references to research since 1981. The present paper uses physical models of this kind to extend the discussion by exploring how aspects of the geometry of the wings may affect their deformations in flight and to suggest how these may in theory be actively influenced and controlled remotely from the axilla. It will focus primarily on species that have either been specifically investigated or for which good photographic information is available. A selection of highspeed photographs by Stephen Dalton and John Brackenbury, some but not all previously published, are included with the authors' generous permission.

I am not concerned here with the shape changes in the expanded anal fans of the hindwings of Orthoptera, Dictyoptera and some other orders (23, 37, 38, 39, 52), or with the flight deformations in Coleoptera hindwings, which are strongly influenced by the flexible lines by which these fold up at rest [17]. The emphasis is on mechanisms involving some transverse bending, particularly, but not exclusively, in forewings. Hindwings also deform in many groups, depending on their relative length and on the presence or absence of wing coupling. The latter also influences the nature of forewing deformation—compare the Trichoptera in Figure 2.

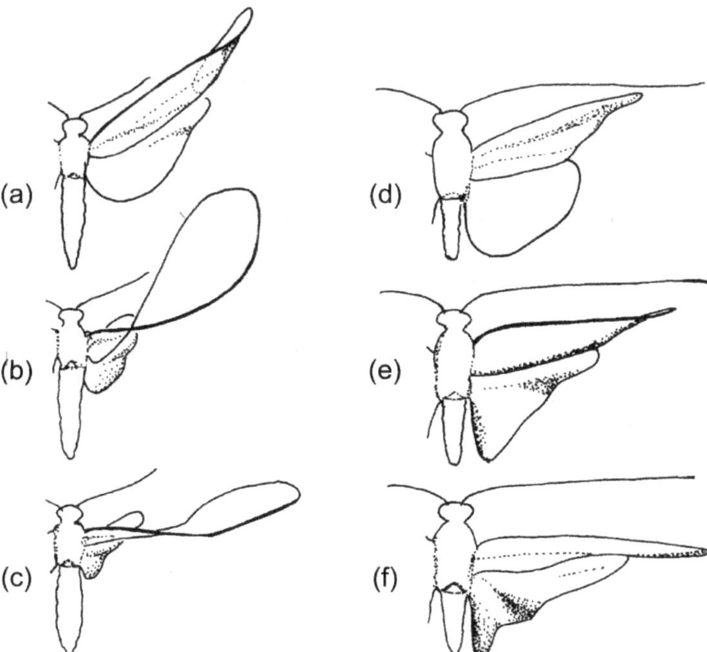

Figure 2. Tracings of successive frames from films of two Trichoptera in tethered flight, comparing a species with uncoupled wings (**a–c**) with a species in which the wings are coupled (**d–f**). From [56], redrawn after [6]. (**a–c**) *Rhyacophila nubile* (Rhyacophilidae). (**d–f**) *Ceraclia senilis* (Leptoceridae).

1.1. Rigidity, Flexibility and Active Control

In typical flight-adapted wings, certain areas are clearly adapted for rigid support, with thick veins, high relief and sometimes thickened membrane. These are generally in the proximal part of the wing and along the more anterior veins. Posterior support, necessary to prevent the wing from pitching into the airflow, is in the forewings and many hindwings of Neoptera generally provided by a rigid clavus, or in many Diptera by automatic mechanisms that lower the trailing edge in response to aerodynamic loading, a situation which is also characteristic of Odonata [12,15,50]. The forewings of Ephemeroptera have no clear clavus, but the anal area provides similar posterior stiffening.

In most insects, the profile of the wing base can alter by hinge-wise bending along specific longitudinal flexion lines [1,2,4,5], of which the most important and widespread are the claval flexion line and the median flexion line—the "remigial furrow" of Martynov [57] and Grodnitsky and Morozov [8]. Basal profile change is a frequent component of the active torsion of the whole wing during the stroke cycle and is the only way in which thoracic muscles can directly deform the wing.

All other deformations are passive responses to aerodynamic, inertial and occasional impact forces, and they tend to be concentrated in more distal areas of the wing, where the relief is flatter, the longitudinal veins are more slender, even sometimes absent, and cross-veins are relatively thin and flexible. These areas are sometimes clearly delineated by a visible transverse flexion line, marked by local areas of thinning of membrane and veins—"thyridia"—or by points or lines of soft cuticle that interrupt the veins themselves.

1.2. The Functions of Bending

Typical wings are thin, springy plates, stiffened by tubular veins, whose mass and thickness diminish along the span. Bending is often a simple response to the inertial forces as the wings decelerate at stroke reversal. Importantly, they only significantly flex ventrally; sometimes around the bottom of the stroke, followed by a sharp straightening, and sometimes throughout most of the upstroke. Dorsal bending is normally slight or absent, though long wings can sometimes flex alarmingly in response to gusts of wind or in extreme accelerations. Otherwise, the principal function of deformation is aerodynamic optimisation: to create necessary force asymmetry between the downstroke and the upstroke, or to generate bursts of unsteady lift.

The shape of the downstroke is fairly consistent: the wing is extended and pronated, usually slightly cambered, with a degree of spanwise twist—"washout", the ideal situation for generating steady lift. Upstroke deformations can be far greater. In some cases, they merely serve to "feather" the wing by reducing its effective area or its angle of attack, so minimising adverse aerodynamic force, but many insects need to develop usefully directed lift throughout the stroke cycle. For this, passive torsion within the span is usually crucial. In Odonata, in many Diptera and in some other insects with uncoupled wings, most of the remigium can swing across like a sail around the supporting anterior veins, but in many other insects—particularly those with coupled wings, like Hemiptera, Hymenoptera, many Lepidoptera and some Trichoptera (Figure 2), or those with a long clavus—torsion is concentrated more distally and is facilitated by a degree of ventral bending, often accompanied by a reversal of camber from dorsally convex to dorsally concave. This brings the distal part of the wing into a favourable angle of attack and suitable profile for generating usefully directed force in the translational part of the stroke, and the dynamic process of changing shape can probably create valuable unsteady lift around stroke reversal.

This paper will use models to investigate the relationships between bending, torsion and camber in wings of this kind, including some Ephemeroptera, Hemiptera, Plecoptera, Megaloptera, Mecoptera and Hymenoptera. These three aspects of deformation are intimately connected. Flexural and torsional rigidity are affected by relief, which in wings can take the form of camber, corrugation or a combination of the two. Whereas a flat plate, or a relatively flat corrugated plate, is equally flexible to dorsal and ventral bending, camber in a thin plate imposes bending asymmetry, as a force applied to the concave side tends to increase the height of the section and hence its rigidity, while a force on the convex side causes the sides to buckle outwards and the section to flatten—an effect familiar to anyone who has used an extending steel ruler [1]. The supporting areas of wings commonly have a degree of built-in camber, ensuring that wing flexion is always ventral. Where the aerodynamic and inertial forces are centred behind the torsional axis, a cambered wing is also asymmetric in twisting, far more resistant to pronation than to supination [51,53] Both these properties are appropriate to the upstroke and are often crucial in determining the shape and attitude of the distal, most aerodynamically effective part of the wing. Under aerodynamic loading, the deformable area of the wing often assumes a cambered section, dorsally convex in the downstroke but concave in the upstroke, and this reversal of camber is also related to aspects of the wing's geometry, as we shall see.

1.3. Modelling Insect Wing Deformation

For the purpose of modelling, I am distinguishing three types of support.

1.3.1. Corrugation

Ephemeroptera and Odonata have fully corrugated wings, with all main vein stems diverging from close to the wing base and alternately occupying the crests and troughs of a fluted structure. Odonate wings do not bend significantly, but in several families of Ephemeroptera, Edmunds and Traver [58] found "bullae"—patches of soft, flexible cuticle—aligned across the wing in three or four of the main concave veins of the forewings, and they correctly identified these as adaptations to ventral bending. Mayfly bullae and their alignment have recently been described in more detail [59].

Ephemera species (Figure 3) have bullae on four major longitudinal concave veins: the subcosta SCP, two branches of the posterior radius RP and the posterior media MP, in a nearly straight line across the wing. Brodsky [19] observed bending in the subimago of *Ephemera vulgata* in flight, though he did not see it in the imago. Four other mayfly families, with quite different flight behaviours, have no bullae. Images found online of the much photographed *Palingenia longicauda*, which does not have bullae, show that ventral flexion can occur in their absence; the bullae appear to be adaptations for sharp, small-radius bending, without damage to the veins.

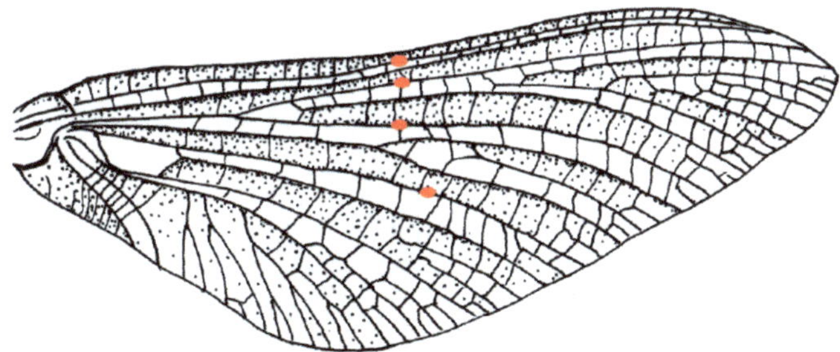

Figure 3. Forewing of *Ephemera vulgata* (Ephemeroptera). The positions of the bullae are shown by red spots.

1.3.2. An M Section

The remigial supporting areas of Plecoptera, Megaloptera, Mecoptera, Trichoptera and many Lepidoptera and Diptera typically have two longitudinal concave troughs. The leading edge spar formed by the costa C, the subcosta SCP and the anterior radius RA is the first; it provides support as far as the point where the SCP ends as a separate vein. The second trough follows the median flexion line, close to the media M in most Plecoptera, *Sialis* (Megaloptera), *Panorpa* (Mecoptera), and most Trichoptera and Diptera. Lepidoptera vary greatly [5]. In Noctuidae, like the *Phlogophora* figured here, the median flexion line lies well anteriorly in the wing. Transverse bending occurs in some members of all these orders, often (except for Diptera) in both fore and hind wings. Wing deformation in Diptera also varies depending on proportions and on the presence or absence of one or more costal breaks and flexion lines [13].

The pattern of bending is strongly influenced by the length of SCP, which often terminates very short of the wing tip, so that the anterior concavity is flattened beyond. Here, and beyond the clavus which provides posterior support, the section is like a shallow letter M or an inverted W.

Figure 4 shows a selection of wings in these groups, together with some photographs and drawings demonstrating the deformations they undergo. The drawings indicate the main flexion lines.

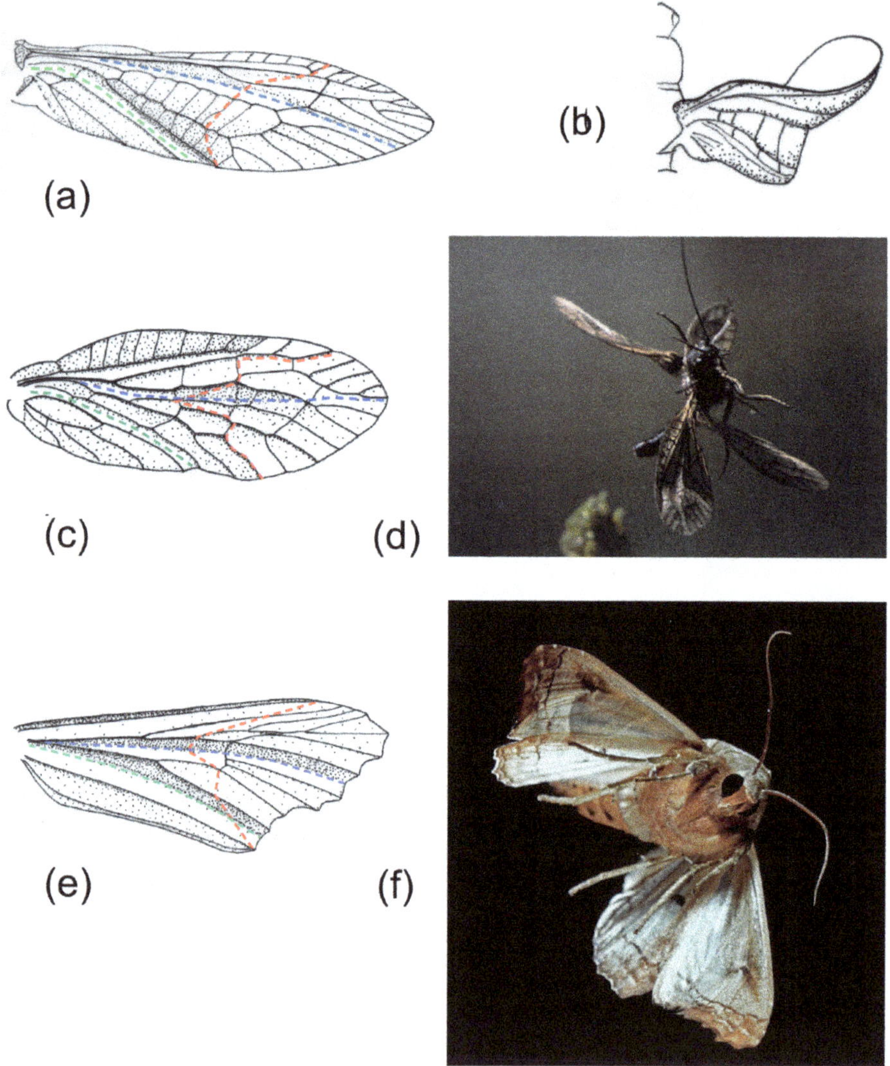

Figure 4. (**a**) *Isogenus nubecola* forewing. (**b**) Tracing of a frame of *I. nubecola* at the start of the upstroke in tethered flight. (**c**) *Sialis lutaria* forewing. (**d**) *S. lutaria* in late upstroke. (**e**) *Phlogophora meticulosa* forewing. (**f**) *P. meticulosa* in late upstroke. (**a**) and (**b**) are redrawn after Brodsky [60]. (**d**) and (**f**) are copyright Stephen Dalton. (**d**) has previously been published in [61], (**f**) in [62]. Red: transverse flexion line. Blue: median flexion line. Green: claval flexion line.

A series of comparative investigations in Russia in the 1980s and 1990s have supplied valuable information on wing deformations in flight [2,5–8,19,21,60]. In all cases, the insects were tethered, so the kinematics may not necessarily reflect free flight, but they illustrate the deformations that the wings allowed. Brodsky [60] filmed *Isogenus nubecula* (Plecoptera) and took a series of high-speed photographs of *Sialis morio* (Megaloptera) [19]. Ivanov filmed *Rhyacophila nubile, Ceraclia senilis, Brachycentrus subnubilis* and *Arctopsyche ladogensis* (Trichoptera) [6], and Grodnitsky with colleagues filmed a range of Lepidoptera [5,8,21]. All showed a degree of ventral bending at the end of the downstroke. Figure 4b, of *Isogenus* immediately after the extreme point of transverse bending, shows a deep groove in the remigium proximal to the flexion, and the distal area is strongly supinated. Later frames from the same sequence show rapid straightening and completion of torsion early in the upstroke, and Brodsky's images of *Sialis* and Ivanov's of Trichoptera show relatively fast recovery, but a high-resolution photograph of *Sialis lutaria* in free flight by Dalton [62] (Figure 4d) shows flexion and reversed camber at an advanced upstroke stage. The same seems to be the case in his photograph of *Phlogophora meticulosa* (Figure 4f), indicating that flexion is maintained throughout the half-stroke, as it appears in Ennos' film of *Panorpa* (Figure 1) and in some of Grodnitsky's moth images [21].

1.3.3. Camber

The basal sections of the forewing remigium of most auchenorrhychous Homoptera have a cambered section. The membrane between the veins is often thickened, a condition that is more strongly developed in the hemielytra of Heteroptera. The camber sometimes continues into the more deformable, distal area; otherwise, this is flat. The clavus, which varies considerably in length, is typically strongly three-dimensional and rigid, and any bending happens at or beyond its apex. A median flexion line is probably frequently present, though not always obvious in Homoptera and detectable only by manipulation [20].

Cicadas (Figure 5a–c) have a particularly obvious transverse flexion line in the forewing, as do Tettigarctidae, Hylicidae and some Cixiidae and Psyllidae [20]. In each case, the line follows a curved path from a break in the costal margin to the end of the clavus, with the apex of the curve towards the wing base—significantly, as we shall see. A variety of other Homoptera show the parallel development of a transverse, straight alignment of cross-veins, presumably localising bending. Cicadas have no median flexion line. Photographs by John Brackenbury (Figure 5b,c) show different degrees of transverse bending and camber reversal in *Tibicina haematodes*.

The hemielytra of Heteroptera (Figure 6) show the clearest differentiation between supporting and deformable areas in any insect. Posterior support continues beyond the clavus as a sclerotised bar at the trailing edge of the remigium. Betts [9,10] found that ventral flexion in flight does not follow the line of the corium margin but takes place within the deformable membrane, along a straight line between the anterior end of the corium and the tip of the posterior sclerotised bar, and this is evident in Brackenbury's photograph of *Palomena prasina* (Pentatomidae) (Figure 6b, from [16]). Several families of Heteroptera have an additional transverse flexion line within the corium: the cuneal fracture. In Miridae, at least, bending can occur at this point as well as at the tip of the corium, increasing the degree of distal supination (Figure 6c, from [16]).

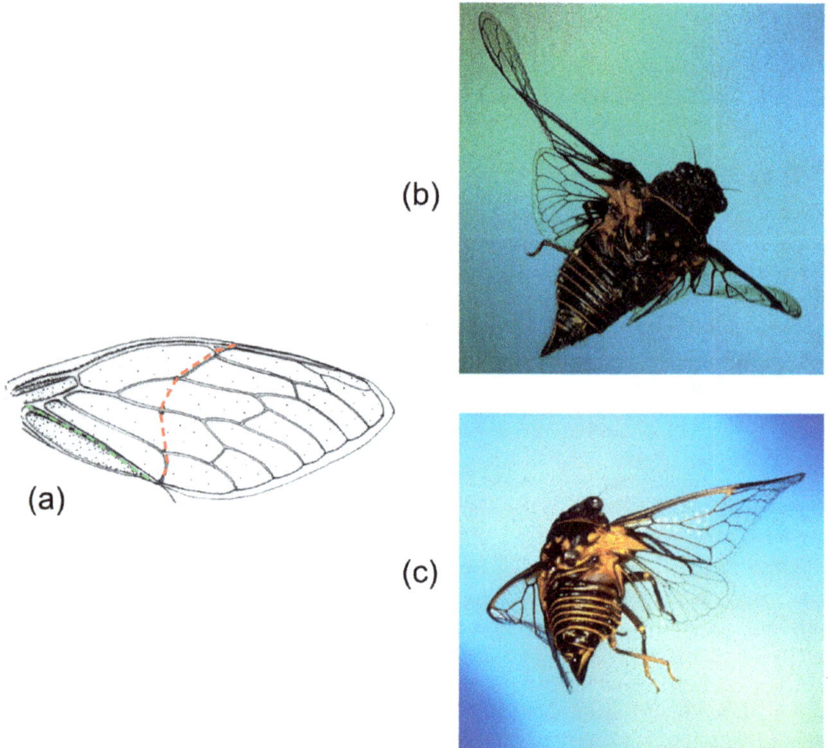

Figure 5. Upstroke deformation in *Tibicina haematodes* (Cicadidae). (**a**) The forewing. (**b**), (**c**) Two images of the upstroke. (**b**) Mid-upstroke, (**c**) early upstroke. (**b**) and (**c**) copyright John Brackenbury. (**b**) has previously been published in [16]. Red: transverse flexion line. Green: claval flexion line.

A cambered wing base is also typical of Hymenoptera, where fusion of the stems of M and the anterior cubitus CUA has eliminated the usual difference in relief between the two veins. Brackenbury [18] has reviewed wing deformation in a range of Hymenoptera. Figure 7b,c, from [16], clearly show flexion, torsion and camber reversal in a wood wasp and an ichneumon, and a photograph of a vespid in [61] and various high speed video sequences which are available online indicate that these are widespread in the order. In coupled wings like these, flexion in the small hindwings is virtually absent, and in-span bending and torsion are restricted to the distal part of the forewing, beyond the coupling.

Examples of forewing M sections and cambered sections are shown in Figure 8. Note that both categories show an overall dorsally convex curvature, ensuring preferential resistance to dorsal bending. The M section wings are distinguished by the presence of a concave branch of the median vein (arrowed), with the median flexion line adjacent.

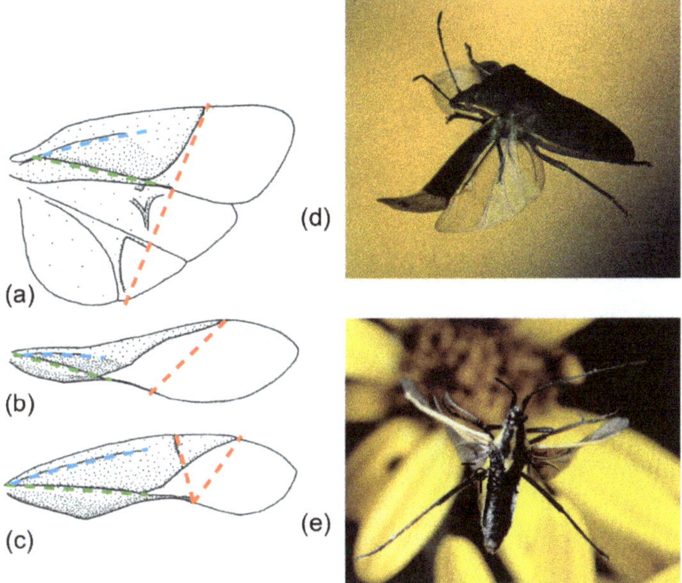

Figure 6. Wing proportions and upstroke deformation in Heteroptera. (**a**) Pentatomidae. (**b**) Alydidae. (**c**) Miridae. (**d**) *Palomena prasina* (Pentatomidae) early in the upstroke. (**e**) *Leptoterna dolabrata* (Miridae) in mid upstroke, showing flexion at the cuneal fracture, aiding supination. (**a**–**c**) From [61], (**d**) and (**e**) copyright John Brackenbury, previously published in [16]. Red: transverse flexion lines. Blue: median flexion line. Green: claval flexion line.

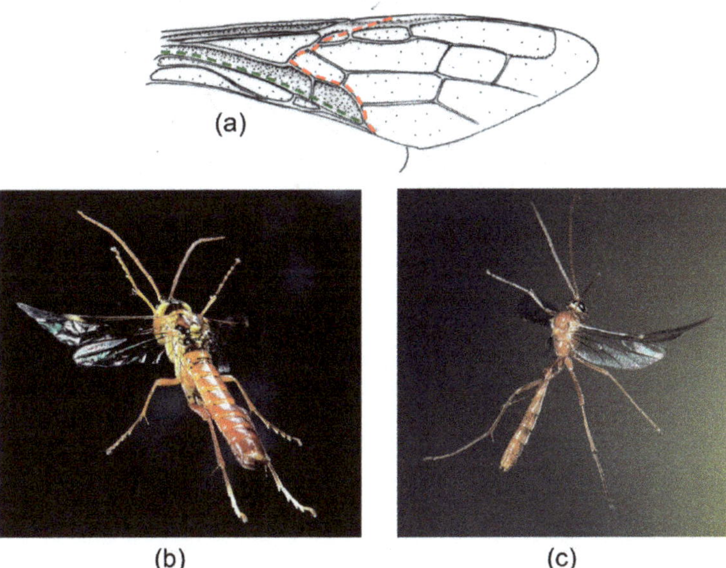

Figure 7. Upstroke deformation in Hymenoptera, showing flexion, torsion and camber reversal. (**a**) Forewing of *Urocerus gigas* (Siricidae). (**b**) *Urocerus gigas* male in mid upstroke. (**c**) *Ophion luteus* (Ichneumonidae) in early upstroke. (**b**) and (**c**) copyright John Brackenbury, previously published in [16]. Red: transverse flexion line. Blue: median flexion line. Green: claval flexion line.

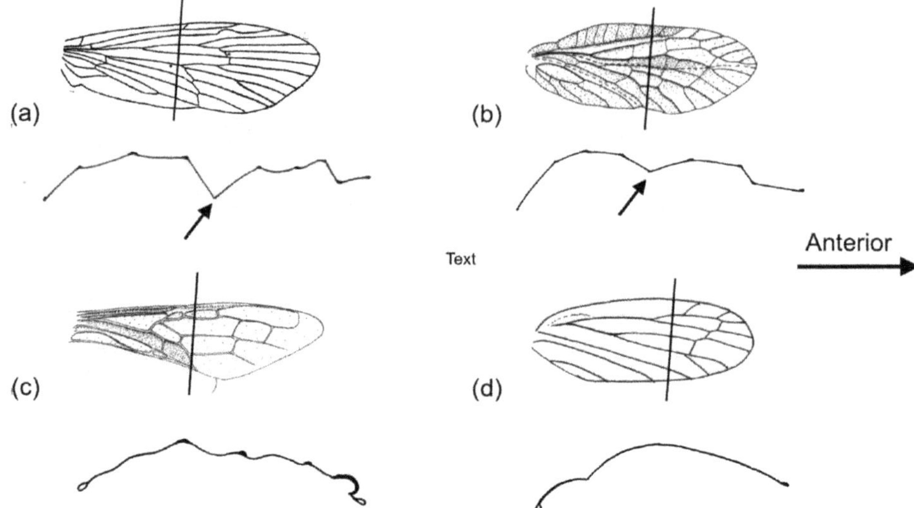

Figure 8. Forewing sections. (**a**) and (**b**) are M sections: (**a**) *Phrygania* (Trichoptera), (**b**) *Sialis* (Megaloptera). (**c**) and (**d**) are cambered sections: (**c**) *Urocerus* (Hymenoptera), (**d**) *Cercopis* (Homoptera). The lines indicate where the sections were cut.

2. Materials and Methods

All models were made of card and paper. The variations in the rigidity and resilience of the different areas of the wing can be crudely replicated by varying the thickness of the materials. The models are simple to construct, and readers are encouraged to make and play with their own versions.

2.1. Corrugation

Model 1 (Figure 9a) was made from an A4 sheet of paper, with a density of 80 g/m^2, and follows Edmunds and Traver [58] in representing the Ephemeroptera condition as a pleated paper fan with a transverse line of notches, simulating the bullae, cut in the concave pleats. With the base held, gentle downward force was applied beyond the bullae until yielding occurred.

2.2. The M Section

Models 2 and 3 (Figure 9b–d) were made from thin card, with a density of 175 g/m^2, though density was not critical. Both measured 29.5 mm × 12 mm. Model 2 represented the M section alone, without anterior support from the leading edge spar or posterior support from the clavus. The sheet was longitudinally folded into three equal panels, and the centre panel was folded in half to form an M section. One end was held firmly by insertion into an expanded polystyrene block, representing the wing base, and downward force was applied to the distal end.

Model 3 had the same dimensions as Model 2, but a long triangular concave fold was added to each of the outer panels, corresponding to the leading edge spar and the clavus, to stiffen the proximal part of the model.

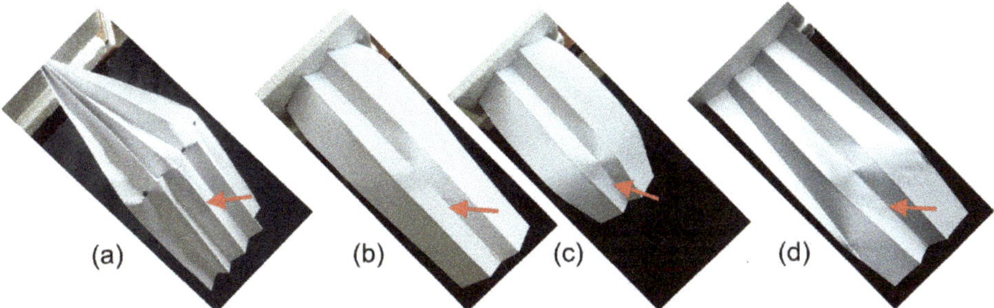

Figure 9. (**a**–**d**) Models 1–3 deforming. (**a**) Model 1. (**b**), (**c**) Model 2. In (**b**), the position is stable. In (**c**), under greater load, the sides are buckling outward, allowing unstable flexion that returns to position (**b**) when the load is removed. Flexion is directly transverse. (**d**) Model 3. The extra anterior and posterior folds delay unstable bending under load, and the flexion line is curved. The arrows indicate the approximate points and directions of the applied bending force.

2.3. Camber

The wings were modelled as rectangles, measuring 29.5 mm × 11.5 mm, with a supporting base made of stiff card and a distal deformable area of standard printing paper, with a density of 80 g/m^2 (Figure 10).

One flexion line AO, parallel to the long sides of the rectangle and corresponding to a median flexion line, was made by cutting partway through the depth of the card. Camber was adjusted experimentally by bending along this line. Its height was represented by the angle ε about the axis AO. The other flexion line, which could be transverse or oblique, was provided by the distal edge of the supporting card. In actual wings, this line is usually curved, but to simplify the geometry in the models, it was made of two straight lines, BO and OD, meeting the median flexion line at point O. This is an acceptable simplification: models with a curved flexion line behaved in exactly the same way.

These models therefore had three variables: the obliqueness of the transverse flexion line, measured by the angle ζ between the longitudinal axis and a straight line joining B and D; the angle BOD, as measured in the flat model; and ε. The first two were part of the model's design, while the third could be manipulated.

In Model 4 (Figure 10a), BOD was straight, so angle BOD = 180° and ζ = 90°. In Model 5 (Figure 10b), BOD = 120° and ζ = 90°. In Model 6 (Figure 10c), BOD = 90° and ζ = 60°. In Model 7 (Figure 10d), the anterior supporting card extended to the end of the model. BOD was 90° and ζ 40°.

One more model, Model 8 (Figure 10e), was produced in order to investigate the specific wing conformation of some families of Heteroptera which have an extra transverse flexion line, the cuneal fracture, and a longitudinal flexion line well anterior to the mid line. The model was made using thinner card than in Models 4-8, as the supporting base needed some flexibility if the model was to function. This is of course closer to the situation in actual insects than the thick card used in the other models; the latter was chosen so that the camber could conveniently be measured as the angle ε.

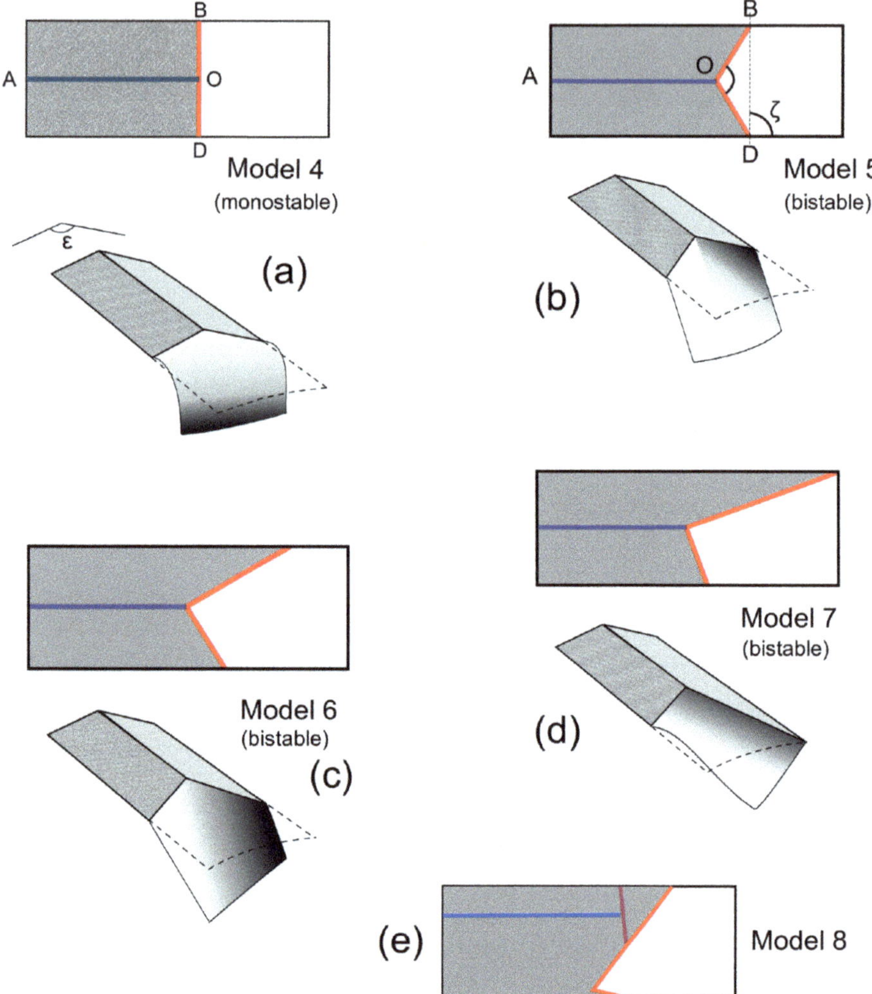

Figure 10. Models 4–8. (**a**) Model 4. (**b**) Model 5. (**c**) Model 6. (**d**) Model 7. (**e**) Model 8. The broken line is the outline in the unflexed state. Model 4 is stable only when unflexed; Models 5, 6 and 7 are bistable. Model 5 shows flexion only; Models 6 and 7 show torsion as well as bending. Model 8: explanation in the text. Red lines correspond to transverse flexion lines, blue lines to median flexion lines in wings.

3. Results

3.1. Corrugation: Model 1

Pressing on the dorsal surface caused the cut pleats to move dorsally and the fan to flatten and bend ventrally, creating an effective one-way hinge. When the fan was allowed to expand laterally, the model was stable only in the unflexed state. In the actual wing, the veins on the ridges and the stiffness of the convex pleats would bring about an elastic return to the unbent state. If expansion was prevented, the fan buckled irreversibly.

3.2. The M Shaped Section, Models 2 and 3

Model 2. Moderate pressure applied to the dorsal side caused the concave ridge to click abruptly upward into the plane of the convex ridges, forming a sharp hinge in the concave ridge, with only minimal curvature in the convex ridges and momentary slight lateral elastic overall expansion, which recovered as the new position was reached; the model was bistable. Further pressure reached a threshold at which the sides of the model buckled outwards and the section underwent catastrophic, unstable bending, returning elastically to the intermediate position when pressure was released.

Model 3. Moderate pressure applied to the dorsal side caused the concave ridge to click upwards, as in Model 2. The extra anterior and posterior folds extended beyond the resulting hinge. A shallow v-shaped flexion line developed between the apices of the extra folds and the hinge. When the dorsal side was pressed harder, the extra folds prevented lateral buckling, and the model was able to undergo appreciably greater stable flexion (Figure 9d). In supplementary models in which the extra folds were shorter, overall bending did occur beyond their apices.

3.3. Cambered Sections, Models 4–9

When flat, and $\varepsilon = 180°$, all models responded equally to forces applied to the upper and lower surfaces, but as soon as slight camber was introduced, they bent only ventrally. Figure 10 illustrates what happened to each model when camber was applied to the card component, slightly reducing ε, and a downward bending force was applied by finger to the flexible paper component distally to the transverse flexion line. The same deformations could be induced by drag forces if the models were flapped.

When camber was applied to the base, Model 4 (Figure 10a), where angle BOD = 180° and $\zeta = 90°$, was stable in only one position, with a positive camber in the paper component. A downward force on the flexible area bent it ventrally, but it returned elastically as soon as the force was released. Other models, not illustrated, in which BOD was 180° and ζ was acute, were also monostable.

All the other models, Models 5–7 (Figure 10b–d), where angle BOD < 180°, had two stable positions: straight, with a positive camber in the paper component, and deflected, with a negative camber. They could be snapped from one position to the other by downward and upward finger pressure. Model 5 simply bent, but in Models 6 and 7, the paper component twisted as well as bent. Other models, not illustrated, showed that the ratio of torsion to bending increases as ζ decreases, and this reached an extreme in Model 8, where the anterior support extended to the end of the model and there was no overall bending.

Models, again not illustrated, where ζ was constant but angle BOD varied, showed that the magnitude of bending and torsion at a given value of ε increased as BOD increased. The geometry here is essentially the same as that described by Haas and Wootton [54] in a practical and theoretical analysis of the mechanisms involved in the folding of the hind wings of beetles and some blaberid cockroaches, and the analytical model which they derived can be applied to the present problem. For this reason, I have used the same letters for points and angles as appear in their paper.

In their model (Figure 11a,b), four fold lines, three of one sense (concave or convex) and one of the other, meet at a single point, the "origin" O. If the model is planar and is capable of being folded completely flat along these lines, opposite pairs of angles around the origin must each total 180°.

In Figure 10, Models 5, 6 and 7 can be seen to correspond to those in Figure 11. In these, when camber was applied by reducing ε, and the paper membrane depressed into the concave position the latter assumed a curved section, whose apex automatically assumed the position of the convex fold line OC in Figure 11a. We can redraw Figure 10c, Model 6, as Figure 11d, with the line of the apex of the curve represented by a line, OC. Figures 10c and 11d are effectively Figure 11c upside down, with angle ε below the model, AO, BO and DO convex instead of concave, and BC concave.

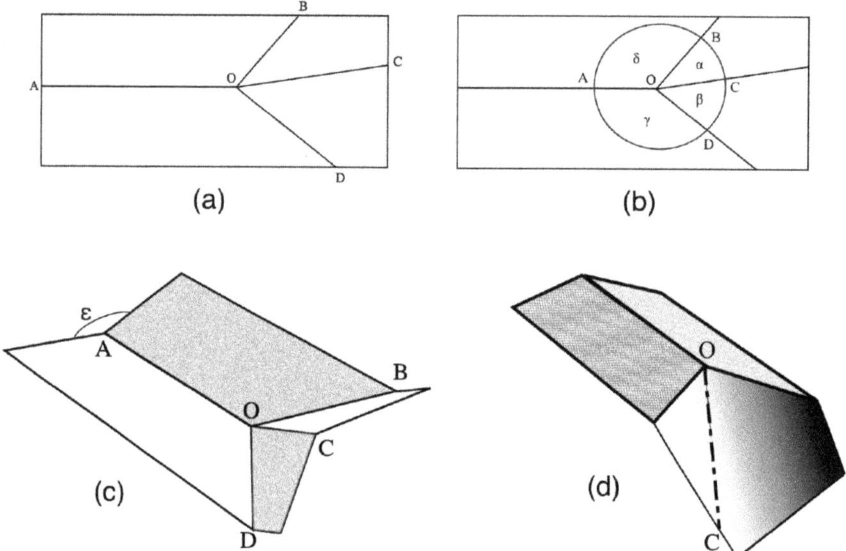

Figure 11. (**a**) A four-fold system, found extensively in Coleoptera hindwings. (**b**) The same, schematized for analysis by Haas and Wootton [54]. (**c**) The system partly folded. (**d**) Figure 9c modified, with a line representing the lowest axis of the curved membrane. Greek letters in (**b**) represent the planar angles around the origin, O.

Haas and Wootton [54] applied vector analysis to calculate the coordinates of point C: c (x), c(y), c(z), for any given value of ε, assuming the fold lines to be of equal length, equal to 1. Using the dot product, they derived three simultaneous equations:

$$\mathrm{Cos}\ \alpha = c(x)^*\cos \delta + c(y)^*\sin \delta \cos \varepsilon + c(z)^*\sin \delta^*\sin \varepsilon$$

$$\mathrm{Cos}\ \beta = c(x)^*\cos \gamma + c(y)^*\sin \gamma$$

$$1 = c(x)^2 + c(y)^2 + c(z)^2$$

Comparing Models 4–7: experimenting by manipulation shows that for a given value of ε, the magnitude and speed of deflection increase and leverage decline with greater values of angle BOD. When BOD is large, a tiny increase in ε causes significant bending, as well as torsion if ζ is acute. In Model 4 (Figure 10a), with BOD = 180°, deflection is theoretically maximal, as the paper could fold back flat over the cardboard—but, in fact, increasing ε merely increases distal camber; there is no leverage to drive deflection.

Manipulating Model 8 showed that increasing the camber about the longitudinal flexion line stiffened anterodistal support, opposing bending at the cuneal fracture.

4. Discussion

The limitations of the models discussed here are self-evident. Insect wings are not rectangles, flexion lines are often not straight or angular, and paper and card do not replicate the gradations in stiffness and resilience of insect cuticle. Their justification lies in the fact that they mimic deformations that are *known* to happen in flight. They are developed by experimenting with materials until their behaviour when manipulated matches that observed in actual wings. They then provide an appropriate first method for examining the geometry and mechanics underlying wing deformations, and they serve to give some direction to future investigations.

Corrugation in wings provides rigidity to transverse bending, but allows compliance to deformation that is parallel to the ridges and channels, and also to torsion provided that any cross-veins are flexible or have flexible joints with the longitudinal veins, like those discovered in Odonata by Newman [3] and comprehensively mapped by Appel and Gorb [63]. Odonata wings show torsion and camber change between half-strokes but have no bending adaptations; any bending being large-radius elastic responses to extreme loads, with immediate recovery.

This is not so with Ephemeroptera. Flight in mayflies, though brief, is crucial to reproductive success, and there is every reason to suppose their wings to be highly adapted for aerodynamic efficiency in the competitive circumstances of mating and oviposition. Bullae are characteristic of the families that use vertical nuptial flights. Vertical flight with a horizontal stroke plane allows aerodynamic force symmetry between the half-strokes, but bending may be needed for force asymmetry in directional flight by the subimagines and females; more kinematic information is needed. Bending requires the wing to flatten, compressing the veins in the concave pleats, and the bullae, like the notches in Model 1, allow these to buckle into the plane of the ridge veins without damage, with the stiffness of the ridge veins driving the elastic return. The bullae are almost in a straight line across the wing, so that flexion will be unstable, and the wing will return elastically, driven by the stiffness of the ridge veins.

The same problem faces other groups that use high relief for rigidity but need to bend. Hemiptera and Hymenoptera tend to meet this by relatively sharp differentiation between the rigid supporting base of the remigium and a flatter distal area, using the properties of camber to limit bending to ventral only. Many other insects, particularly among Plecoptera and Holometabola, show more gradual diminution of relief along the span and allow bending across moderate relief by upward buckling. Here, this is true of only one vein and an adjacent flexion line but is similar in principle to the situation in Ephemeroptera and paralleled in Models 2 and 3. In these cases, thyridia often serve the same function as the bullae of mayflies, allowing local buckling without damage.

With both solutions, there seems to be a distinction between some situations, as shown in Brodsky's film of *Isogenia*, where bending—sometimes extreme—takes place around stroke reversal, followed by rapid straightening and torsion in the early part of the upstroke, and others, where some flexion continues throughout the upstroke, usually combined with some torsion and camber reversal. Both are likely to have aerodynamic consequences. The sharp, angular acceleration in the former situation may create useful transient unsteady lift; the latter condition would give steady favourable lift throughout the translational part of the upstroke.

In the former case, monostable bending, as in Models 3 and 4, is acceptable, but in the second, bistability is useful, and a curved flexion line is common, as simulated in Models 5, 6 and 7 and visible in those of *Sialis*, *Phlogophora*, *Panorpa*, *Tibicina*, *Urocerus* and *Ophion* (Figures 5–7). In these cases, bending can contribute to torsion, and an oblique flexion line becomes valuable—expressed as ζ in Models 6 and 7, where flexion and torsion are interdependent. The inclination ζ depends greatly on the relative lengths of the anterior and posterior supports—of SCP and the clavus (in Heteroptera, the secondary rigid extension). This is illustrated in the Heteroptera in Figure 6a, and in the difference between the fore and hind wings of *Panorpa* (Figure 1) [14]. The ratio of bending to torsion depends on the value of ζ. The extreme condition, with torsion only, occurs where the anterior support extends to the wing tip, as in Model 7, e.g., in Odonata, sphingid moths and many Diptera and Hymenoptera, although many flies and Hymenoptera have a costal break—two in some Diptera—which allow a degree of flexion [18]. This can also enhance torsion, and the same is true of the cuneal fracture in mirid bugs (Figure 6c). Costal breaks in many Diptera are unusually proximally situated, and ventral flexion and consequent torsion are sometimes extreme—for example, in the supremely kinematically versatile *Calliphora*, whose wing can flex at two points, namely at the end of the SCP and close to the base [1,64,65]. Ennos [65] has discussed the possible aerodynamic implications of this, suggesting that the option of ventral flexion may give extra control of the force vector in all planes and contribute to their remarkable manoeuvrability.

Very little change in the basal camber of Model 7 was needed to alter distal wing torsion significantly, and the same was true of both bending and torsion in Model 6. Model 8 is also significant. As Betts [9] showed, the cuneal fracture may offer the options of flexion there, or across the membrane, or both, and this could well be controllable by altering the basal section about the median flexion line.

The physical models described demonstrate some mechanisms by which insects could potentially remotely control the instantaneous rigidity and shape of their wings in flight—but do they? The basal section in many insects certainly alters during the stroke by hinge-wise bending along the claval flexion line, but it is not yet clear how often and to what extent flexion along the median flexion line is actively employed to influence distal shape and attitude in flight. Basal camber could in theory be modified by altering the timing and/or amplitude of shortening of the basalar and subalar muscles. These typically act antagonistically to pronate and supinate the wing respectively over the fulcrum of the pleural wing process; a reduction, phasic or tonic, in the shortening amplitude of one or both could potentially induce camber in part of the stroke, but it may not be as simple as this. In the well-documented case of locust forewings, which control the distal angle of attack by assuming a basal z-shaped profile in the upstroke by flexion about both the median and claval flexion lines [66,67], the basalar and subalar muscles apparently act together to pronate the wing in the downstroke, while the principal supinator is the flexor muscle [68]. Heteroptera, many of which have a clear median flexion line in the corium that would seem to make them excellent candidates for active section control, have no basalar muscles; the wing is pronated phasically by the indirect dorsal longitudinal muscle acting through the first and second axillary sclerites [9,69]. The subalar muscle could perhaps induce basal camber by shortening tonically over several stroke cycles, but this is pure conjecture.

Too little is still known about the precise operation of the basal direct muscles and axillary sclerites of most insects, and electrophysiological as well as morphological research will be necessary to determine whether in any particular case of stroke-by-stroke variation in wing shape is actively controlled and wing rigidity actively tuned.

Whether or not the insects exert active profile control, the mechanisms do have possible technical applications. Much recent work has gone into designing wings for micro air vehicles, but these have for the most part been relatively unsophisticated, utilising the wings' flexibility but not attempting section control. I have suggested elsewhere how the principles explored in this paper could be used in an insect-based MAV, with minimal additional actuation [70].

5. Conclusions

Interest in the intricate, fascinating structure of insect wings has grown enormously in recent years, with the expansion of biomimetic engineering and the development of new micromorphological techniques and computational modelling. Understandably, the emphasis has been on a few species, predominantly Odonata and Diptera, with outstanding flight capabilities. The broader picture provided by comparative studies, and hence of interest to entomologists as well as engineers, has in general been lacking. This paper has attempted to show how simple, quickly built, physical models can continue to be useful in investigating aspects of wing design, in explaining parallel adaptations across the range of insect groups and by indicating directions for more sophisticated modelling.

Funding: This research received no external funding.

Acknowledgments: I acknowledge with thanks the permission of the following societies for permission to reproduce the following figures: the Royal Entomological Society for Figure 1b, the Norwegian Academy of Science and Letters for Figure 2 and the Entomological Society of America for Figure 6a–c. I am particularly grateful to Stephen Dalton and John Brackenbury for allowing me to crop and reproduce their superb high-speed photographs, in Figure 4 (S.D.) and Figures 5–7 (J.B.).

Conflicts of Interest: The author declares no conflict of interest.

References

1. Wootton, R.J. Support and deformability in insect wings. *J. Zool.* **1981**, *193*, 447–468. [CrossRef]
2. Antonova, O.A.; Brodsky, A.K.; Ivanov, V.D. Wingbeat kinematics in five insect species. *Zool. Zhurn.* **1981**, *60*, 506–518. (In Russian)
3. Newman, D.J.S. The functional wing morphology of some Odonata. Ph.D. Thesis, University of Exeter, Exeter, UK, 1982.
4. Brodsky, A.K.; Ivanov, V.D. The functional assessment of insect wings. *Entomol. Obozr.* **1983**, *62*, 241–256. (In Russian)
5. Grodnitsky, D.L.; Kozlov, M.V. The functional morphology of the wing apparatus and peculiarities of flight of primitive moths. *Zool. Zhurn.* **1985**, *64*, 1661–1671. (In Russian)
6. Ivanov, V.D. Comparative analysis of wingbeat kinematics of caddisflies (Trichoptera). *Entomol. Obozr.* **1985**, *64*, 273–284. (In Russian)
7. Brodsky, A.K. Structure and functional significance of veins and furrows in insect wings. In *Morphological Bases of Insect Phylogeny*; Nauka: Leningrad, Russia, 1987; pp. 4–19. (In Russian)
8. Grodnitsky, D.L.; Morozov, P.P. Morphology, flight kinematics and deformation of the wings in holometabolous insects (Insects: Oligoneoptera = Scarabaeiformes). *Russ. Entomol. J.* **1994**, *3*, 3–32.
9. Betts, C.R. Functioning of the wings and axillary sclerites of Heteroptera during flight. *J. Zool. Lond. (B)* **1986**, *1*, 283–301. [CrossRef]
10. Betts, C.R. The kinematics of Heteroptera in free flight. *J. Zool. Lond. (B)* **1986**, *1*, 303–315. [CrossRef]
11. Wootton, R.J.; Betts, C.R. Homology and function in the wings of Heteroptera. *Syst. Entomol.* **1986**, *11*, 389–400. [CrossRef]
12. Ennos, A.R. The importance of torsion in the design of insect wings. *J. Exp. Biol.* **1988**, *140*, 137–160.
13. Ennos, A.R. Comparative functional morphology of the wings of Diptera. *Zool. J. Linn. Soc.* **1989**, *96*, 27–47. [CrossRef]
14. Ennos, A.R.; Wootton, R.J. Functional wing morphology and mechanics of *Panorpa germanica* (Insecta: Mecoptera). *J. Exp. Biol.* **1989**, *143*, 267–284.
15. Wootton, R.J. Functional morphology of insect wings. *Annu. Rev. Entomol.* **1992**, *37*, 113–140. [CrossRef]
16. Brackenbury, J. *Insects in Flight*; Blandford: London, UK, 1992; pp. 1–192.
17. Brackenbury, J. Wing folding and free flight kinematics in Coleoptera (insecta): A comparative study. *J. Zool.* **1994**, *232*, 253–283. [CrossRef]
18. Brackenbury, J. Hymenopteran wing kinematics: A qualitative study. *J. Zool. Lond.* **1994**, *233*, 523–540. [CrossRef]
19. Brodsky, A.K. *The Evolution of Insect Flight 1994*; Oxford University Press: Oxford, UK, 1992; pp. 1–299.
20. Wootton, R.J. Functional wing morphology in Hemiptera systematics. In Studies in Hemipteran Phylogeny. Schaefer, C.W., Ed. Entomological Society of America: Lanham, MD, USA, 1996; pp. 179–198.
21. Grodnitsky, D.L. *Form and Function in Insect Wings. The Evolution of Biological Structures*; Johns Hopkins University Press: Baltimore, MD, USA; London, UK, 1999; pp. 1–261.
22. Dudley, R. *The Biomechanics of Insect Flight*; Princeton University Press: Princeton, NJ, USA; Oxford, UK, 2000; pp. 1–474.
23. Wootton, R.J.; Evans, K.E.; Herbert, R.C.; Smith, C.W. The hind wing of the desert locust (*Schistocerca gregaria* Forskal). 1. Functional morphology and mode of operation. *J. Exp. Biol.* **2000**, *203*, 2921–2931.
24. Wootton, R.J.; Herbert, R.C.; Young, P.G.; Evans, K.E. Approaches to the structural modelling of insect wings. *Phil. Trans. R. Soc. Lond. B* **2003**, *358*, 1377–1587. [CrossRef] [PubMed]
25. Combes, S.A.; Daniel, T.L. Flexural stiffness in insect wings. I. Scaling and the influence of wing venation. *J. Exp. Biol.* **2003**, *206*, 2979–2987. [CrossRef] [PubMed]
26. Combes, S.A.; Daniel, T.L. Flexural stiffness in insect wings. II. Spatial distribution and dynamic wing bending. *J. Exp. Biol.* **2003**, *206*, 2989–2997. [CrossRef]
27. Ellington, C.P. The aerodynamics of hovering insect flight. III. Kinematics. *Philos. Trans. R. Soc. Lond. B* **1984**, *305*, 41–78.
28. Ellington, C.P. The aerodynamics of hovering insect flight. IV. Aerodynamic mechanisms. *Philos. Trans. R. Soc. Lond. B* **1984**, *305*, 79–113.

29. Willmott, A.P.; Ellington, C.P.; Thomas, A.L.R. Flow visualization and unsteady aerodynamics in the flight of the hawkmoth *Manduca sexta*. *Philos. Trans. R. Soc. Lond. B* **1997**, *352*, 306–316. [CrossRef]

30. Young, J.; Walker, S.M.; Bomphrey, R.; Taylor, G.K.; Thomas, A.L. Details of insect wing design and deformation enhance aerodynamic function and flight efficiency. *Science* **2009**, *325*, 1549–1552. [CrossRef]

31. Sane, S.P. The aerodynamics of insect flight. *J. Exp. Biol.* **2003**, *206*, 4191–4208. [CrossRef] [PubMed]

32. Wang, Z.J. Dissecting insect flight. *Annu. Rev. Fluid Mech.* **2005**, *37*, 183–210. [CrossRef]

33. Zhao, L.; Huang, Q.; Deng, X.; Sane, S.P. Aerodynamic effects of flexibility in flapping wings. *J. R. Soc. Interface* **2009**, *7*, 485–497. [CrossRef]

34. Mountcastle, A.M.; Daniel, T.M. Aerodynamic and functional consequences of wing deformation. In *Animal Locomotion*; Taylor, G.K., Triantfyllou, M.S., Tropea, C., Eds.; Springer: Heidelberg, Germany, 2010; pp. 311–320. [CrossRef]

35. Mountcastle, A.M.; Combes, S.A. Wing flexibility enhances load-lifting capacity in bumblebees. *Proc. R. Soc. Lond. B* **2013**, *280*, 20130531. [CrossRef] [PubMed]

36. Nakata, T.; Liu, H. Aerodynamic performance of a hovering hawkmoth with flexible wings: A computational approach. *Proc. R. Soc. Lond. B* **2011**, *279*, 722–731. [CrossRef]

37. Smith, C.W.; Herbert, R.C.; Wootton, R.J.; Evans, K.E. The hind wing of the desert locust (*Schistocerca gregaria* Forskal). II. Mechanical properties and functioning of the membrane. *J. Exp. Biol.* **2000**, *203*, 2933–2943.

38. Herbert, R.C.; Young, P.G.; Smith, C.W.; Wootton, R.J.; Evans, K.E. The hind wing of the desert locust (*Schistocerca gregaria* Forskal). III. A finite element analysis of a deformable structure. *J. Exp. Biol.* **2000**, *203*, 2945–2955.

39. Walker, S.M.; Thomas, L.R.; Taylor, G.K. Deformable wing kinematics in the desert locust: How and why do camber, twist and topography vary through the stroke? *J. R. Soc. Interface* **2008**, *6*, 735–747. [CrossRef] [PubMed]

40. Du, G.; Sun, M. Effects of wing deformation on aerodynamic forces in hovering hoverflies. *J. Exp. Biol.* **2010**, *213*, 2273–2283. [CrossRef] [PubMed]

41. Walker, S.M.; Thomas, A.L.R.; Taylor, G.K. Deformable wing kinematics in free-flying hoverflies. *J. R. Soc. Interface* **2010**, *7*, 131–142. [CrossRef]

42. Tanaka, H.; Whitney, J.P.; Wood, R.J. Effect of flexural and torsional wing flexibility on lift generation in hoverfly flight. *Integr. Comp. Biol.* **2011**, *51*, 142–150. [CrossRef]

43. Wootton, R.J.; Newman, D.J.S. Evolution, diversification and mechanics of dragonfly wings. In *Dragonflies and Damselflies. Model Organisms for Ecological and Evolutionary Research*; Cordoba-Aguila, A., Ed.; Oxford University Press: Oxford, UK; New York, NY, USA, 2008; pp. 261–275.

44. Donaughe, S.; Crall, J.D.; Merz, R.A.; Combes, S.A. Resilin in dragonfly and damselfly wings and its implication for wing flexibility. *J. Morphol.* **2011**, *272*, 1409–1421. [CrossRef]

45. Rajabi, H.; Darvizeh, A. Experimental investigations into the functional morphology of dragonfly wings. *Chin. Phys. B* **2013**, *22*, 088702. [CrossRef]

46. Bomphrey, R.J.; Nakata, T.; Henningsson, P.; Lin, H.T. Flight of the dragonflies and damselflies. *Philos. Trans. R. Soc. Lond. B* **2016**, *371*. [CrossRef]

47. Rajabi, H.; Gorb, S.N. How do dragonfly wings work? A brief guide to functional roles of wing structural components. *Int. J. Odonatol.* **2020**, *23*, 23–30. [CrossRef]

48. Nakata, T.; Henningsson, P.; Lin, H.T.; Bomphrey, R.J. Recent progress on the flight of dragonflies and damselflies. *Int. J. Odonatol.* **2020**, *23*, 41–49. [CrossRef]

49. Newman, D.J.S.; Wootton, R.J. An approach to the mechanics of pleating in dragonfly wings. *J. Exp. Biol.* **1986**, *125*, 361–362.

50. Wootton, R.J. The functional morphology of the wings of Odonata. *Adv. Odonatol.* **1991**, *5*, 153–169.

51. Wootton, R.J. Leading edge section and asymmetric twisting in the wings of flying butterflies (Insecta, Papilionoidea). *J. Exp. Biol.* **1993**, *180*, 105–118.

52. Wootton, R.J. Geometry and mechanics of insect hindwing fans: A modelling approach. *Proc. R. Soc. Lond. B* **1995**, *262*, 181–187.

53. Ennos, A.R. Mechanical behavior in torsion of insect wings, blades of grass and other cambered structures. *Proc. R. Soc. Lond. B* **1995**, *259*, 15–18.

54. Haas, F.; Wootton, R.J. Two basic mechanisms in insect wing folding. *Proc. R. Soc. Lond. B* **1985**, *263*, 1631–1658.

55. Wootton, R.J. Invertebrate paraxial locomotory appendages: Design, deformation and control. *J. Exp. Biol.* **1999**, *202*, 3333–3345.
56. Wootton, R.J. Design, function and evolution in the wings of holometabolous insects. *Zool. Scripta.* **2002**, *31*, 31–40. [CrossRef]
57. Martynov, A.V. On the two types of insect wings and their evolution. *Russ. Zool. Zhurn.* **1924**, *4*, 155–185. (In Russian)
58. Edmunds, G.F.; Traver, J.R. The flight mechanics and evolution of the wings of Ephemeroptera, with notes on the archetype insect wing. *J. Wash. Acad. Sci.* **1954**, *44*, 390–400.
59. Dominguez, E.; Abdala, V. Morphology and evolution of the wing bullae in South American Leptophlebiidae (Ephemeroptera). *J. Morphol.* **2019**, *280*, 95–102. [CrossRef] [PubMed]
60. Brodsky, A.K. The evolution of the flight apparatus of stone flies (Plecoptera). Part 3. In-flight wing deformation in stone fly *Isogenus nubecula* Newman. *Entomol. Obozr.* **1981**, *60*, 523–534.
61. Dalton, S. *Borne on the Wind*; Chatto and Windus: London, UK, 1975; pp. 1–160.
62. Dalton, S.; Bailey, J. *At the Water's Edge*; Century Hutchinson: London, UK, 1989; pp. 1–160.
63. Appel, E.; Gorb, S.N. Comparative functional morphology of vein joints in Odonata. *Zoologica* **2014**, *159*, 1–104.
64. Nachtigall, W. *Insects in Flight*; Allen and Unwin Ltd: London, UK, 1974; pp. 1–153.
65. Ennos, A.R. The kinematics and aerodynamics of the free flight of some Diptera. *J. Exp. Biol.* **1989**, *142*, 49–85.
66. Jensen, M. Biology and physics of locust flight. III. The aerodynamics of locust flight. *Philos. Trans. R. Soc. Lond. B* **1956**, *239*, 511–552.
67. Wootton, R.J. Function, homology and terminology in insect wings. *Syst. Entom.* **1979**, *4*, 81–93. [CrossRef]
68. Pfau, H.K. Zur Morphologie und Funktion des Vorderflügels und Vorderflügelgelenks von *Locusta migratoria* L. *Fortschr. Zool.* **1977**, *24*, 341–345.
69. Govind, C.K.; Dandy, J.W.T. The thoracic mechanism of the milkweed bug, *Oncopeltus fasciatus* (Heteroptera, Lygaeidae). *Can. Entomol.* **1970**, *102*, 1057–1074. [CrossRef]
70. Wootton, R.J. Springy shells, pliant plates and minimal motors: Abstracting the insect thorax to drive a micro-air vehicle. In *Flying Insects and Robots*; Floreano, D., Zufferey, J.-C., Srinivasan, M.V., Ellington, C.P., Eds.; Springer: Heidelberg, Germany, 2009; pp. 207–217. [CrossRef]

 *insects*

Review

Wing Design in Flies: Properties and Aerodynamic Function

Swathi Krishna, Moonsung Cho, Henja-Niniane Wehmann, Thomas Engels and Fritz-Olaf Lehmann *

Department of Animal Physiology, Institute of Biosciences, University of Rostock, 18059 Rostock, Germany; swathi.krishna@uni-rostock.de (S.K.); moonsung.cho@uni-rostock.de (M.C.); henja-niniane.wehmann@uni-rostock.de (H.-N.W.); thomas.engels@uni-rostock.de (T.E.)
* Correspondence: fritz.lehmann@uni-rostock.de; Tel.: +49-381-498-6301

Received: 21 June 2020; Accepted: 19 July 2020; Published: 23 July 2020

Abstract: The shape and function of insect wings tremendously vary between insect species. This review is engaged in how wing design determines the aerodynamic mechanisms with which wings produce an air momentum for body weight support and flight control. We work out the tradeoffs associated with aerodynamic key parameters such as vortex development and lift production, and link the various components of wing structure to flight power requirements and propulsion efficiency. A comparison between rectangular, ideal-shaped and natural-shaped wings shows the benefits and detriments of various wing shapes for gliding and flapping flight. The review expands on the function of three-dimensional wing structure, on the specific role of wing corrugation for vortex trapping and lift enhancement, and on the aerodynamic significance of wing flexibility for flight and body posture control. The presented comparison is mainly concerned with wings of flies because these animals serve as model systems for both sensorimotor integration and aerial propulsion in several areas of biology and engineering.

Keywords: locomotion; animal flight; wing structure; aerodynamics; flight force

---

1. Introduction

Insect wings are complex, three-dimensional structures that are under selective pressures towards functional optima. These optima result from multiple requirements, and also from evolutionary influences relevant to the animal's fitness. Wings have mainly evolved for locomotion and produce aerodynamic forces during gliding and flapping flight at high wing beat frequencies of up to 1000 Hz [1]. The air flows generated for flight mainly depend on wing kinematics, the wing's overall planform, and the dynamics of elastic deformation owing to inertial and aerodynamic loading. Pinpointing the factors that shape the evolution of wings and flapping kinematics is key to any in-depth understanding of flight. Within the past decades, numerous comprehensive reviews and book chapters have been published on insect flight, focusing on components such as aerodynamic mechanisms for lift enhancement [2–11], power requirements for wing flapping [12–15], wing kinematics and control [16–21], and the efficiency with which muscle mechanical power is turned into weight supporting lift [22,23]. This review is engaged in the link between three-dimensional wing structure and aerodynamics, focusing on recently published studies on the aerodynamic performance of wings in differently-sized insects. The review highlights the behavior of wings in flies because these animals often serve as model systems for aerial propulsion in both biology and engineering.

Insect wings receive their mechanical strength and endurance from two main components: on the microscopic level, the three-dimensional composition of proteins and chitin-based cuticle layers [24–27], and on the macroscopic level, the distribution and three-dimensional morphology of veins and elastic interconnecting membranes [28–33]. This light-weight design helps insect wings to widely resist external forces using chitin as the main chemical component [34]. Veins greatly vary in density, size, and shape between animal species and determine the wing's structure and mechanical behaviors under load, such as bending and twisting [29,35–39]. Veins provide structural support to a wing, preventing the wing from tear [40,41] and host sensory receptors such as campaniform sensilla and innervated bristles, including their afferent nerves [42–47]. By contrast, wing membranes are aerodynamic active surfaces and composed of multiple layers of cuticle [25,27,48] with a thickness ranging from ~0.5 μm in small insects to ~1.0 mm in forewings (elytra) of large beetles [28,49]. Veins and membranes form fine geometrical structures that are typically of much smaller scale than the primary flow structures at wings, such as wing tip and leading edge vortices, and referenced as wing corrugation [50]. Coarse-scale structures, by contrast, typically refer to the wing's overall curvature and termed chordwise and spanwise wing camber [51]. Throughout the past decades, several technical developments, such as high-resolution micro-computed tomography (μCT), have helped to better understand the various aspects of wing morphology for structural integrity [27,52], while robotic and numerical studies on insect flight have highlighted the aerodynamic significance of three-dimensional wing design [53–57].

Numerous studies have been published on the aerodynamic performance of translating [58–66] and root-flapping rigid wings [8,67–74]. The aerodynamics of dynamically deforming insect wings, by contrast, is less clear. Wing bending and twisting change the wing's local angle of attack during flapping motion. Wing bending and twist is thus similar to changes in wing kinematics and change flow and force production. Wings may have an anisotropy in mean stiffness for ventral versus dorsal loading that unbalances force production during upstroke and downstroke, even in cases in which wing hinge articulation is the same in both halfstrokes [27,37,75]. Moreover, as spanwise stiffness in insect wings is approximately one to two orders of magnitude larger than chordwise stiffness, wings often deform in a characteristic fashion [37,76]. There is a continuing debate on the potential benefits of dynamic shape changes in flapping flight because some authors reported aerodynamic advantages of wing deformation for lift production [77–80], while other authors found disadvantages [80–82].

In this review, we work out the significance and tradeoffs of wing design for aerodynamic key parameters such as vortex development and lift production. This is achieved by disassembling the wing's various properties and linking the components in wing structure to aerodynamics, power consumption and flight efficiency. The sections start with flow phenomena in a simple, flat, rectangular wing. In the second section, we focus on the benefits of elliptical and tapered wing shapes as found in many species, including flies. This section also highlights that even simple genetic modifications of fly wing planforms lead to measurable changes in aerodynamic performance. In the third section, we consider the wing's three-dimensional morphology. A recent numerical study, for example, showed that the three-dimensional shape of rigid fly wings attenuates both lift production and aerodynamic efficiency rather than enhancing these measures compared to a flat wing [83]. In the last section, we focus on the aerodynamic consequences of elastic deformation in morphological complex wings. Although elastic wings share similar fluid dynamic properties with rigid wing, an animal must cope with the dynamically changing conditions because these changes may attenuate the ability and precision of flight and body posture control.

2. Aerodynamic Properties of Root-Flapping Rectangular Wings

Rigid, flat, rectangular wings are often used to understand fundamental aerodynamic principles and represent the most simple approach towards insect flight [84] (Figure 1). They are investigated at different kinematic patterns such as revolving [85–87] and pitching motions [88–91]. Most studies though focused on the dynamics of the leading edge vortex that develops on the upper wing side at high angle of attack [8,72,92–100]. In contrast to a translating wing at high Reynolds number, the leading edge vortex in root-flapping and revolving insect wings is stably attached to the dorsal wing surface and enhances lift throughout the stroke cycle [72,92,101]. It obtains its stability from the viscosity of air and axial flow between wing hinge and wing tip [102,103]. Although a rectangular root-flapping plate produces all characteristic types of vortices and flows typical for insect wings, it suffers from low span efficiency compared to an elliptically shaped insect wing. Span efficiency is similar to Rankine–Froude efficiency, which typically refers to mean efficiency of propulsion in a complete wing flapping cycle of an animal during hovering conditions [104,105]. By contrast, instantaneous span efficiency varies during wing flapping and is the ratio between ideal power requirements for lift production and the actual requirements [106]. Span efficiency is maximum when the distribution of vertical velocities is uniform in the wing's downwash [107,108]. Under this condition, the kinetic energy of the downwash is minimal owing to the non-linear, velocity-squared relationship between kinetic energy and wake velocity. If velocities vary within the wake, the velocity-squared relationship produces costs at elevated velocities that are not saved by the regions with low fluid velocities (Figure 2).

A pair of translating, flat wings has maximum span efficiency if it produces an elliptical lift distribution from tip to tip (Figure 2b) [109]. Span efficiency depends on the geometry of a wing, i.e., planform and camber, and its kinematics, but not on the wing's aspect ratio and wing loading [107]. In general, the left and right wing of a two-winged insect can either be considered a single aerodynamic system or both wings may function as two aerodynamically independent systems. In the first case, each wing should have a semi-elliptical shape that results in an ellipse if both wings are connected via the insect body, where as in the second case each wing should have an elliptical shape for maximum span efficiency. Both geometrical cases yield higher span efficiency than a translating rectangular wing with same aspect ratio, and are thus beneficial for gliding flight of an insect. However, this conclusion only holds if the wings are flat and not twisted because an appropriate twist of a rectangular wing may equalize the downwash distribution via changes in local angle of attack.

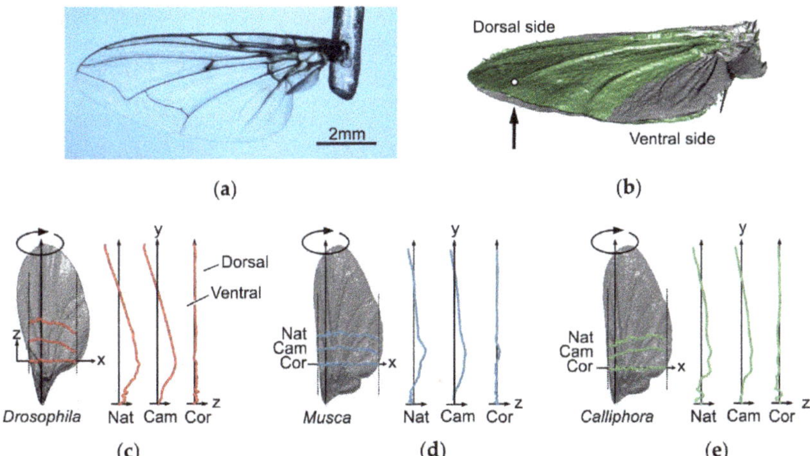

Figure 1. Characteristics of fly wings. (**a**) Detached wing of the blowfly *Calliphora vomitoria*, mounted to a steel holder. (**b**) Deformation of a blowfly wing (green) during loading by a ~64 µN point force (white dot) applied normal to the ventral wing side (arrow) [75]. Grey, surface profile without load. (**c–e**) Spanwise and chordwise wing profiles along the axes of rotation in three differently-sized fly species (*Drosophila melanogaster, Musca domestica, Calliphora vomitoria*). The wing profiles are superimposed on natural wing models (grey). The profiles separately show wing camber (Cam) and wing corrugation (Cor). Both wing components were numerically extracted from the natural wing shape (Nat) according to a procedure outlined in Engels et al. [83]. The out-of-plane component (z) is exaggerated by a factor of 2 for better clarity.

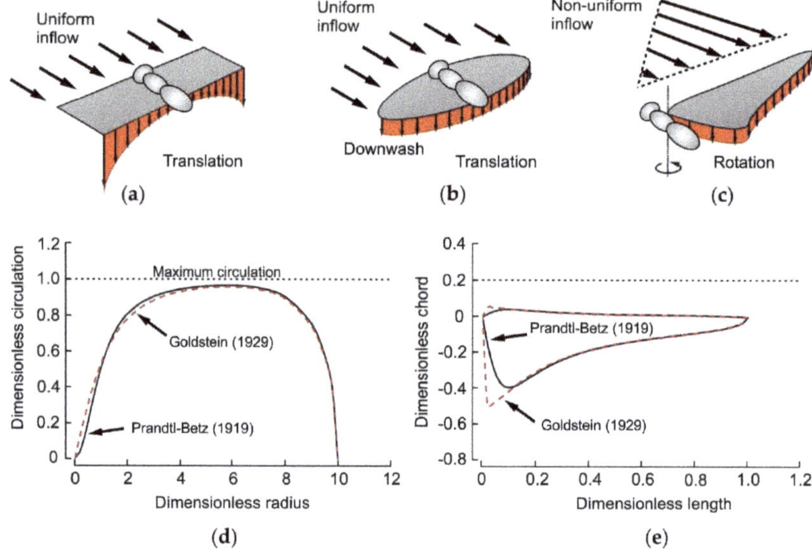

Figure 2. Ideal distribution of spanwise lift in translating and revolving wings. Distribution of vertical downwash velocity during translation in an (**a**) rectangular and (**b**) elliptical insect wing. At constant forward flight velocity, the inflow towards the wing is uniform. The ideal elliptical wing shape spreads spanwise vorticity that produces maximum span and Rankine–Froude efficiencies. (**c**) In a revolving wing, the non-uniform inflow requires adjustments in wing shape for maximum efficiency. (**d**) Distribution of spanwise circulation in an elliptical wing according to Prandtl [109], Betz [110] and Goldstein [111]. (**e**) Ideal wing shape for maximum span efficiency in a revolving wing according to Prandtl–Betz and Goldstein (see Supplementary Materials).

In contrast to translating wings, in revolving and root-flapping wings, local blade velocity increases with increasing distance from root to tip, producing a non-uniform inflow distribution (Figure 2c). This changes the ideal, root-to-tip elliptical distribution in circulation (Figure 2d). Thus, an elliptical wing does not produce a uniform downwash distribution during revolving or root-flapping motion, requiring an eccentric planform for maximum span efficiency. Betz, Prandtl and Goldstein [110–112] estimated the optimal distribution of circulation in flat propeller wings, assuming flow leakages at the tip and root and thus zero circulation at the revolving axis (Figure 2c). Based on their results, we estimated the optimal wing shape in Figure 2e and for the calculations in Figure 3 (see Supplementary Materials). In contrast to Betz and Prandtl, Nabawy and Crowther [113–115] derived the optimal wing shape of two revolving wings assuming the elliptical circulation distribution of a pair of translating wings, with maximum circulation at the revolving axis. In this theoretical case, wing chord must continuously increase from wing tip to root in order to compensate for the drop in inflow velocity, leading to an "optimum" wing shape [114,115]. However, the latter design cannot produce a uniform downwash as in Prandtl–Betz's estimate. In sum, the expected lower span efficiency in a rectangular wing may have fueled the evolution of elliptical insect wings for gliding flight. The expected lower span efficiency of elliptical wings during wing flapping, by contrast, might have led to the development of wing shapes that taper off towards the wing tip. Besides numerous biological pressures on wing planform development, it should be noted that span efficiency is only one aerodynamic factor that determines the costs of wing flapping as other costs such as inertial power requirements may also significantly contribute to total flight power expenditures [116].

3. The Aerodynamic Benefits of an Ideal Planform

Wing shape in insects is diverse. Significant shape measures are aspect ratio and the wing's planform. High aspect ratio wings minimize induced drag and provide high lift-to-drag ratios by reducing the three-dimensional flow effects associated with tip vortices [117]. Aspect ratio also determines the stability of the leading edge vortex during wing flapping [117]. There is a wide variety of aspect ratios found in insect wings ranging from approximately 1.5 to 5.8 [118–122]. In Diptera, previous studies reported aspect ratios of 2.91–3.14 for *Drosophila* [121,122], 2.88 for *Musca* [83], and 2.62–2.93 for *Calliphora* [119,121]. The highest aerodynamic forces in hovering, root-flapping insect-like wings are produced at an aspect ratio of approximately 3.0 [123]. As already mentioned, wing planform determines both the ability of a wing to produce lift and the span efficiency. Span efficiency for a gliding wing typically varies between 0.7 and 0.85 [106] and previous studies on animal locomotion thus used a standard generic value of 0.83 [108]. The latter value is comparatively close to the maximum efficiency of an ideal wing with elliptical shape for translation and is not reached for root flapping wings at low advance ratios.

Flow measurements in differently-sized moths, for example, show that span efficiency in flapping flight is much smaller and varies between species. As the tested moth species had wings with similar aspect ratio and planform, there is no trend in span efficiency with increasing body size [108]. Lowest efficiency of 0.31 was measured in the smallest moth species *Hemaris fuciformis* with 0.2 g body mass, 0.6 in the intermediate-sized species *Deilephila elpenor* and with 0.85 g body mass and 0.46 in the largest species *Manduca sexta* with 1.44 g body mass [108]. These data imply that the generic value of 0.83 might not be a suitable approximation in flying insects. Eventually, butterfly wing planforms, in particular, produce elevated lift and thrust coefficients compared to any other planforms [124]. In these species, the coefficients of force production increase with increasing taper ratio and aspect ratio. This increasing performance, however, occurs at the cost of increasing power requirements for flight and thus at the cost of a reduction in aerodynamic efficiency [124].

For this review, we additionally calculated the aerodynamic quantities of revolving (Figure 3) and flapping (Figure 4) wings of a blowfly, as well as simple rectangular and ideal-shaped wings in order to compare their performance. The ideal wing shape was calculated according to the estimation by Prandtl–Betz in Figure 2e. The numerical simulations were performed using a previously published numerical model [83,125] combined with a wavelet-adaptive solver [126], and efficiency was calculated as Rankine–Froude efficiency [127]. Table 1 shows that revolving rectangular and fly wings perform similarly, producing approximately the same amount of lift. The fly wing, however, produces this force at slightly higher efficiency (0.23) compared to a rectangular wing (0.22). Both values are approximately half of the values calculated from quasi-steady approach on flapping insects wings [128]. Surprisingly, an ideal-shaped wing for rotation is less effective because most wing area is concentrated at the wing base where the wing's inflow velocity is low. The ideal-shaped wing produces ~52% less lift at ~29% less efficiency than rectangular and natural fly wings (Table 1).

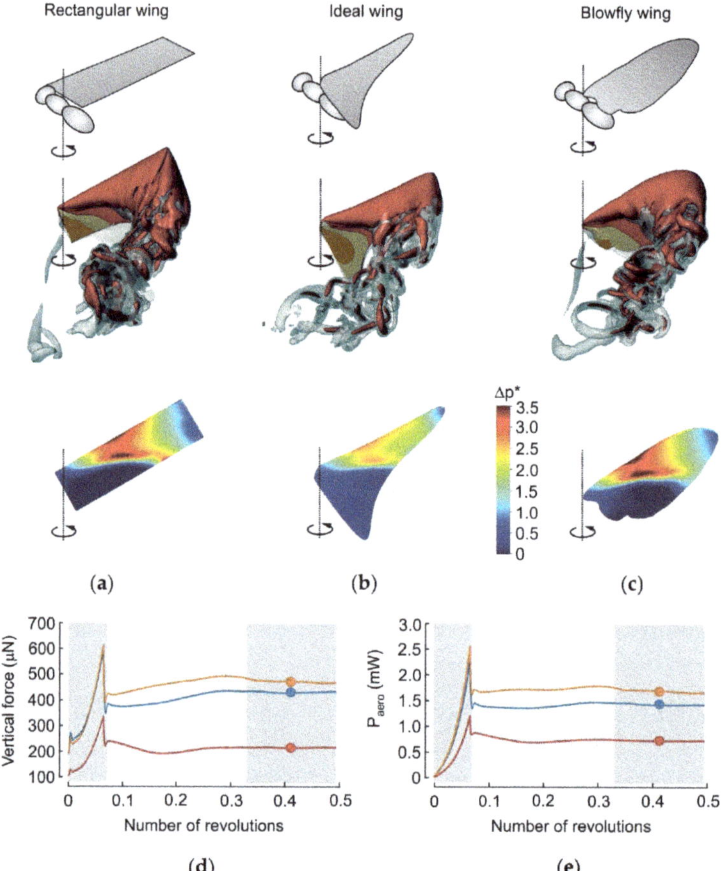

Figure 3. Aerodynamics of revolving wings. (**a–c**) Upper row: aerodynamic characteristics of three flat, continuously revolving wings (rectangular wing, ideal wing for rotation, wing of a blowfly). Middle row: data show iso-surface with vorticity magnitude of 75 s^{-1} (grey) superimposed on a vorticity iso-surface with 150 s^{-1} (red). The flow is shown after ~0.4 revolutions after motion onset. Lower row: pressure difference (Δp *) between dorsal and ventral wing sides, and normalized to the uniform wing loading pressure. The latter value is equal to body weight divided by the surface area of two wings. (**d,e**) Time evolution of vertical lift in *d* and aerodynamic power in *e*. After motion onset (grey, left), lift and power stabilize approximately after 0.3 revolutions (grey, right). Dots are mean values calculated from ~0.32–~0.5 revolutions (grey, right). Wing length and area are identical in all wings. For numerical modeling see [83]. Orange, rectangular wing; blue, wing of *Calliphora vomitoria*; and red, ideal-shaped wing.

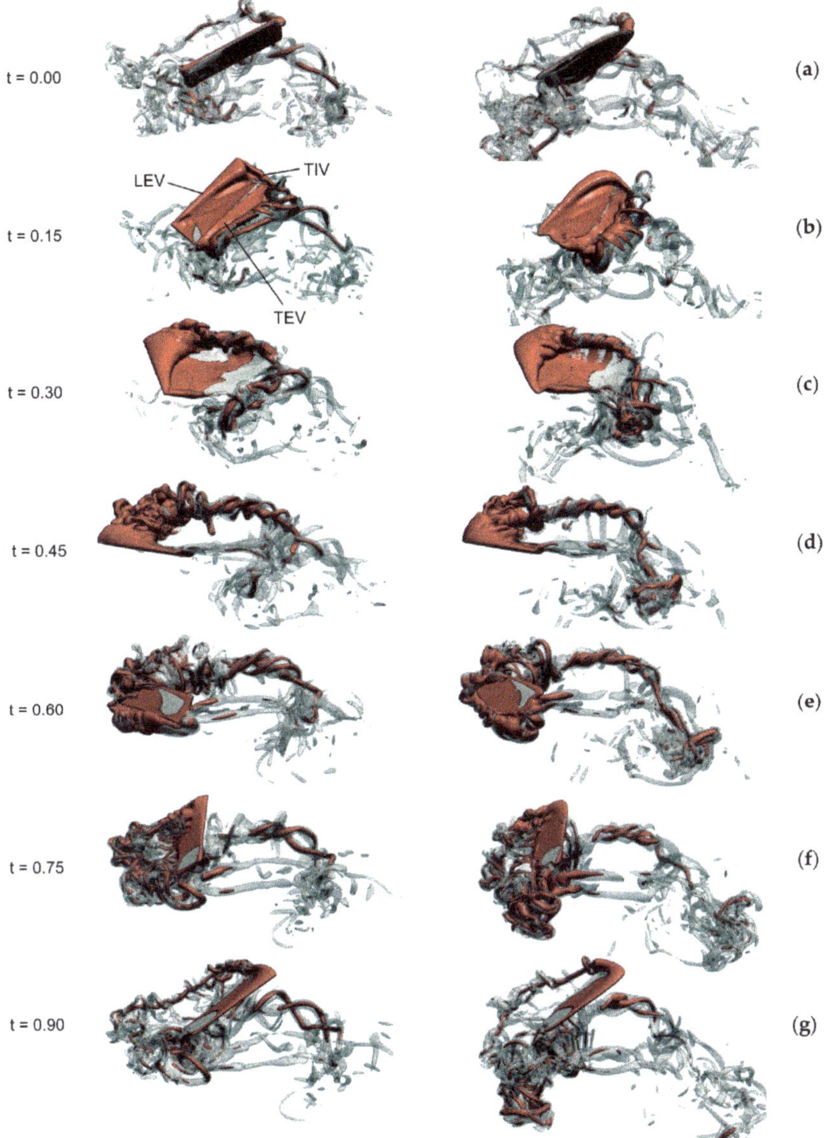

Figure 4. Evolution of vorticity in a flapping rectangular (left) and blowfly (right) wing. (**a–g**) Vorticity distribution at the beginning of the 3rd flapping cycle (t = 0–1) after motion onset. Vorticity of a flapping wing of *Calliphora vomitoria* slightly differs from the flow in the rectangular wing. Data show iso-surface with vorticity magnitude of 75 s^{-1} (semi-transparent grey) superimposed on a vorticity iso-surface with 150 s^{-1} (red). LEV, leading edge vortex; TEV, trailing edge vortex; TIV, wing tip vortex. For performance data and wing kinematics confer to Table 1 and a previously published study [83], respectively. Wing length and area are identical in both wings.

Table 1. Aerodynamic characteristics of single wings with various shape during revolving and flapping motion. Wing shapes are shown in Figures 1–3.

Kinematics	Property	Rectangular Wing	Ideal Wing	Fly Wing
Revolving [1]	Vertical force (μN)	471	215	431
Revolving [1]	P_{aero} (μW)	1696	724	1434
Revolving [1]	Efficiency	0.22	0.16	0.23
Flapping [2]	Vertical force (μN)	479	n.a.	458
Flapping [2]	P_{aero} (μW)	2340	n.a.	2361
Flapping [2]	Efficiency	0.27	n.a.	0.25

Data are calculated by a three-dimensional numerical simulation model that was refined from a previously published code (https://arxiv.org/abs/1912.05371). All tested wings have similar area (28.0 mm^2) and length (9.76 mm), and were flat without corrugation and camber. Mean vertical force was derived from t = ~0.32 to t = 0.5 revolutions after motion onset in the revolving wing, and from the 3rd flapping cycle in flapping wings. Efficiency, Froude efficiency for wing flapping [127]; n.a., no data available. Reynolds number is calculated from mean wing tip velocity and mean wing chord. [1] Horizontal stroke plane, 112 Hz, 40° angle of attack, Reynolds number = 1320. [2] Inclined stroke plane (−20°, nose-down), 40° angle of attack during upstroke, 20° angle of attack during downstroke, 0.22 cycle for wing rotation, 150 Hz stroke frequency, Reynolds number = 1320 [83].

Adding kinematic reversals to the revolving kinematic pattern (flapping motion) has little effect on the performance of a rectangular and natural fly wing (Table 1). However, the time evolution of lift production suggests that a rectangular wing produces more lift during up- and downstroke than the fly wing, while the fly wing produces more lift during the stroke reversals.

Although aerodynamic force production changes with changing wing planform, there is little variation in the wake behind wings with different geometry [129,130] (Figure 4). This is demonstrated by the pressure distribution of differently-shaped wings in Figure 3 and by experimental investigations on different categories of elliptic wing planforms with same aspect ratio and total area at Reynolds numbers typical for wing motion in flying insects between 160 and 3200 [130]. The latter study suggests that wake structure mainly depends on shape of the wing's leading edge rather than planform. The authors argue that the leading edge shape determines the shear layer feeding the leading edge vortex, and thus the development of leading edge vortices and the associated flow topology [131]. Similar results are reported on mosquito flight using computational fluid mechanics and in vivo flow measurements [94]. The latter study shows that apart from leading edge vortices, also trailing edge vortices and rotational drag are responsible for elevated lift production. This was concluded from the low-pressure distribution on the suction side of the wing near the trailing wing edge. The wing planform of fruit flies, by contrast, does not produce similar low pressure regions although both insects fly at similar Reynolds numbers [94].

In general, researchers often assume that the specific wing shape of an insect species is close to an optimum, reflecting the result of a selection process on the animal's aerial performance. A unique approach toward the aerodynamic consequences of wing planforms in flies, however, implies that wing shape also results from aerodynamically non-adaptive factors [47]. Flight tests on fruit flies with genetically modified wing shape using targeted RNA interference demonstrate that wildtype controls, with wing aspect ratios of ~2.5, have a reduced flight capacity compared to transgene animals with wings at aspect ratios between ~2.7 and ~3.0 [47]. While maximum forward flight speed does not increase with increasing aspect ratio, the transgene flies exhibit ~22% improved tangential acceleration and an ~10% improved deceleration capacity, they turned at higher angular rate (~10 – ~21%) and at an ~23% smaller turning radius than controls. The results suggest that in fruit flies, an increasing aspect ratio leads to an increase in agility and maneuverability. Notably, even if the GAL4-induced RNA interference selectively tackled wing shape, the above findings could also be explained by behavioral modifications because the maximum mechanical power output of the indirect flight muscles were thought to be similar in both tested groups [47].

4. Functional Relevance of Three-Dimensional Wing Shape

There is a longstanding debate on the functional relevance of three-dimensional wing shape compared to a flat wing design. It is widely accepted that the wing's three-dimensional corrugation serves as a mechanical design element to improve stiffness and thus to avoid excessive wing deformation during flight [28,132–134]. Its potential contribution to aerodynamic lift and drag production, by contrast, is less clear and apparently depends on the chosen approach for analysis. The majority of previously published studies used numerical or physical wing models at various Reynolds numbers for analysis and reported that wing corrugation either improves aerodynamic performance [56,58,65,66,118,135–137] or attenuates performance [56,59,65,66,134,137–139]. Other studies that reported little or no effect of corrugation on wing performance in beetles [55], dragonflies [140], bumblebees [141], hoverflies [142], and fruit flies [50] at Reynolds numbers between 35 and 34,000. Some studies, moreover, also reported inconsistent results on the significance of wing corrugation in dragonflies [63,64,143,144], bumblebees [54], and a generic model [59].

Although corrugation may change local wing pressure, the difference of lift and drag coefficients between corrugated and flat wings is typically not more than 5% for both lift and drag for angles of attack between 35° and 50° [141], and 17% for drag at low Reynolds number of 200 and 5° angle of attack [50]. A likely explanation for the latter findings is that corrugation is usually smaller than the typical flow structures at the wing, such as the leading-edge vortex and the area of flow separation. Thus small-scale corrugation produces only small local changes in both flows at the wing and aerodynamic forces [50]. As the size of flow structures depends on Reynolds number, corrugation structures should be coarser in small insect wings than in larger wings for pronounced wing-vortex interaction. In contrast to small-scale corrugation, large-scale chordwise wing camber has a pronounced effect on aerodynamics characteristics of a wing [54,55]. Upward camber and a downward oriented leading wing edge tend to create more lift than a flat wing flapping at similar angle of attack. Chordwise camber and the shape of the leading edge are thus comparable to a change in the effective angle of attack of an insect wing [56,83].

There is little difference in flow patterns between flat and three-dimensional fly wings but vortices and stagnant air cushions that are trapped in corrugation valleys of a wing may potentially improve lift production by changes in wing's effective geometry [61,135]. Evidence for trapped vortices were experimentally found in wings moving at relatively high Reynolds number [63,64], including an aerodynamic study that demonstrated vortex trapping at the wing's acceleration phase and at Reynolds numbers ranging from 34,000 to 10^5, but not at 3500 [140]. The latter value is at the upper end of Reynolds numbers typical for flying insects. Studies that did not find vortex trapping attributed the absence to the elevated angle of attack in insect wings [55]. In corrugated wings of gliding dragonflies, slowly rotating vortices only develop at small angles of attack but flow broadly separates from the wing surface at larger angles (Re = 34,000 [53], Re = 1400 [136]). By contrast, a recent numerical study on root-flapping wings shows that corrugation valleys in fruit flies, house flies, and blowflies are unable to trap vortices at Reynolds numbers up to 1623 (Figure 5) [83]. Thus, small-scale corrugation, low Reynolds number, spanwise flow advecting vorticity and high angle of attack make vortex trapping less likely in flapping insect wings. Trapped flows should thus be considered as an exception rather than a common aerodynamic phenomenon in insect flight [134].

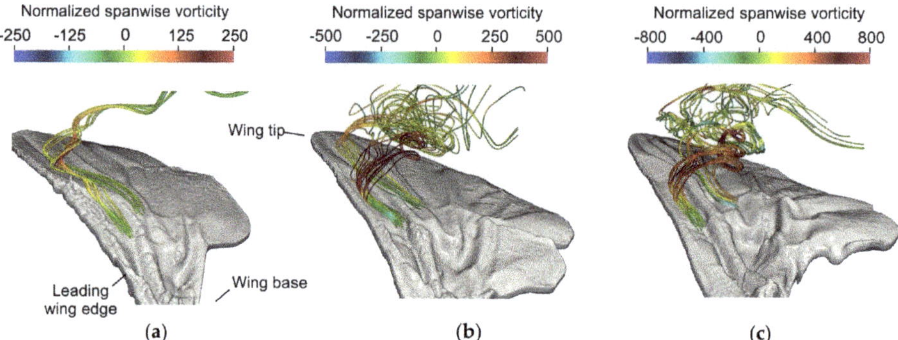

Figure 5. Flow pattern produced by natural wing models of three fly species. Color-coded instantaneous streamlines in (**a**) *Drosophila*, (**b**) *Musca*, and (**c**) *Calliphora*. Snapshots are taken at 1.3 (*Drosophila*) and 3.3 stroke cycle (*Musca, Calliphora*) after motion onset in natural wings [83]. Streamlines were computed from particles released in the corrugation valleys of the dorsal (upper) wing surface near the leading wing edge. Data show little spanwise vorticity inside the corrugation valley near the surface (arrows) and leading-edge vortex suction pulls the virtual particles away from the surface.

Aerodynamic studies on fly wings with genetically modified corrugation and camber patterns are missing and thus is the exact significance of wing corrugation in flies for aerodynamic performance and efficiency. This difficulty was recently circumvented by a numerical study using computational fluid dynamics on three differently-sized fly species (*Drosophila melanogaster*, *Musca domestica*, and *Calliphora vomitoria*) [83]. The wing models were reconstructed from high-resolution scans [75] and corrugation and camber numerically removed afterwards. The study allowed a direct comparison of air flow structures, force production, power requirements, and propulsion efficiency of a natural, cambered, corrugated and flat wing design. The findings suggest that three-dimensional corrugation of fly wings has no significant effect on mean aerodynamic force production compared to a flat wing at the tested Reynolds numbers for wing motion between 137 and 1623 [83]. This result is consistent with a previous study on bumblebee model wings that reported less than 5% change in aerodynamic force production of four differently-corrugated wings [141]. Our data, instead, suggest that corrugation may alter the temporal distribution of forces within the stroke cycle.

The three-dimensional camber of rigid fruit fly-, housefly-, and blowfly-wings also has no significant benefit for lift production but attenuates Rankine-Froude flight efficiency by up to ~12% compared to a flat wing [83]. This is different from previous findings on deforming wings in hoverflies, which is discussed in chapter 5 [145]. The computed flight efficiencies in rigid wings of 17–23% were somewhat below the experimentally derived estimates that range from 26–32% in various species of fruit flies to 37–55% in large crane flies, beetles and bees [23]. A potential explanation for this discrepancy is that many of the experimental studies used Ellington's quasi-steady model for flight power [128], while the numerical model solved the Navier–Stokes equations for fluid motion. Altogether, the above results make it more likely that 3-dimensional corrugation and camber have been selected according to mechanical rather than aerodynamic constraints. Even though there are some energetic costs for wing flapping associated with three-dimensional wing shape, the increased stiffness and change in force distribution in corrugated and cambered insect wings might be of advantage during elevated wing loading—conditions that occur during maneuvering and flight under turbulent environmental conditions.

5. Wing Stiffness and Benefits of Elastic Wing Deformation

Wing joints, the cuticular composition of proteins and chitin fibers, and elastic proteins such as resilin allow wings to elastically deform during flapping motion in response to inertial and aerodynamic loads [24,146–156]. Elastic wing deformation alters flight in two ways: first, it smooths out and thus lowers sudden acceleration of local wing mass, and consequently maximum instantaneous inertial costs [116,157–159], and second, it changes flow conditions due to changes in local angle of attack, and thus the direction of flow [79,160]. In hoverflies, these effects appear to be negligible, as the time courses of lift, drag and aerodynamic power are similar in deforming (camber deformation, spanwise twisting) and rigid flat-plate wings [145]. Part of the potential energy stored in a deformed wing might not be elastically recycled throughout the stroke cycle, which results in plastic deformations and stress on the cuticle [161]. Moreover, the energy loss stresses the total energy budget for flight and thus leads to a reduction of propulsion efficiency. Measurements in wings of fruit flies, house flies and blowflies suggest that only 77–80% [161] and 87–93% [75] of the elastic potential energy is recycled during a full deformation–relaxing cycle. However, the significance of the relative loss in elastic potential energy depends on how much the wing deforms during flight. For example, at the end of each half stroke, aerodynamic and added mass reaction force partly cancel out wing mass-induced moments [161]. Total elastic potential energy is thus small at the end of upstroke and downstroke, and so is energy loss. Consequently, the elastic structures of the wing may not be able to recycle much kinetic energy gained from a preceding half stroke and thus contribute only slightly to the recycling of kinetic energy at the stroke reversals. By contrast, a larger amount of elastic potential energy is stored at the beginning of each half stroke and subsequently released throughout the wing translation phase in flies [161].

To avoid wing bending at elevated wing loading, spring and flexural stiffness of insect wings typically increase with increasing body size [29]. This finding also holds for fruit flies, house flies and blowflies, in which median spring stiffness along an aerodynamic characteristic beamline is ~0.024, 0.63, and 1.76 Nm^{-1}, and median flexural stiffness is 4.86×10^{-11}, 9.73×10^{-9}, and 1.33×10^{-7} Nm^2, respectively [75]. Due to these elevated stiffness values, fly wings deform only little in spanwise direction during wing flapping. Nevertheless, the distribution of local spatial stiffness in fly wings varies between species. In response to point loads at 11 characteristic points on the wing surface, for example, the average spring stiffness of bending lines between wing hinge and point load varies ~77-fold in fruit flies and ~44-fold in house flies but only ~28-fold in large blowflies [75]. This suggests that wings of larger flies behave more like a homogenous material with uniform thickness compared to smaller flies. As this property determines how inertial and aerodynamic forces deform a flapping wing, the stiffness variability could reflect the differences in local aerodynamic forces in different species.

Besides elastic energy recycling, dynamic deformations in span- and chordwise direction alter the wing's aerodynamic performance throughout the stroke cycle [162–164] and may help to stabilize flight [165]. Findings on the aerodynamics of flexible wings have recently been summarized in a comprehensive review [5]. For example, Du and Sun [145] found that camber deforming and spanwise twisting wings of hoverflies produce ~10% more lift at ~17% less aerodynamic power expenditures than a flat rigid wing. The authors suggest that this benefit in lift production is mainly caused by the dynamic changes in wing camber, while the difference in power is mainly due to spanwise twist [145]. More lift at reduced costs results in an increase in flight efficiency, which in turn reduces the metabolic cost for wing flapping and may eventually enhance the animal's fitness. Notably, this conclusion runs counter to the study on rigid fly wings that found a decrease in Rankine–Froude efficiency in cambered compared to flat wings (see chapter 4) [83]. Other examples on the significance of dynamic camber and spanwise twist include beetles and moths. Owing to force-induced deformation, wing camber in beetles is inverted (downward camber) during the upstroke that improves aerodynamic performance compared to a non-deforming wing [55]. Aerodynamic details of wings with different geometry including twist, leading edge details, and camber in hawkmoth-like revolving wings [86] show that flow separation at the leading edge prevents leading-edge suction and thus allows a simple geometric relationship between forces and angle of attack. The force coefficients in these experiments appear to be remarkably invariant against alterations in leading-edge detail, twist and camber. In general, our knowledge on the aerodynamic significance of three-dimensional wing structure and flexing in insect flight is still limited and largely stems from studies on simplified flight models such as two-dimensional computational simulations, rectangular flat wing planforms, simplified three-dimensional extrusions of two-dimensional profiles, and also from work at inappropriately large Reynolds number [54,58,59,65,118,135,136,138].

6. Conclusions

In conclusion, wings of insects and wings of flies (in particular) are complex, three-dimensional body appendages with elevated spanwise and comparatively little chordwise stiffness. Their tapered shape improves span efficiency during root-flapping but genetic modifications of wing shape has questioned that the current shape solely results from a evolutionary selection process towards maximum aerodynamic performance [47]. The three-dimensional corrugation pattern of veins and membranes forms valleys that channel axial flow components, following the pressure gradient from the wing hinge to the tip, but does not trap vortices for lift-enhancement as previously suggested for the more corrugated wings of dragonflies [28,61,83,135]. Fly wings also have the ability to store elastic potential energy during wing deformation, but analyses using static loadings suggest that up to ~20% of this energy might be lost due to plastic or viscoelastic deformation. Nevertheless, the exact benefits of three-dimensional wing design for locomotor capacity, flight efficiency and body posture control in insects are still under debate [166]. These data, however, are highly welcome not only by biologists working on insect flight, but also by engineers working in the area of bionic propulsion and on the development of the next generation of man-made flapping devices.

Supplementary Materials: Detailed descriptions on the calculation of spanwise circulation and wing shape in revolving wings are available online at http://www.mdpi.com/2075-4450/11/8/466/s1.

Author Contributions: S.K. prepared an early draft of the review; M.C. calculated the ideal wing profile for a rotating wing; T.E. performed the numerical calculations; H.-N.W. revised the draft and prepared references; F.-O.L. improved the manuscript. All authors equally contributed to editing of this review. All authors have read and agreed to the published version of the manuscript.

Funding: This research received no external funding.

Acknowledgments: The authors thank the referees for their helpful comments on this manuscript.

Conflicts of Interest: The authors declare no conflicts of interest.

References

1. Sotavalta, O. The flight-tone (wing stroke frequency) of insects. *Acta Entomologica Fenn.* **1947**, *4*, 1–117.
2. Chin, D.D.; Lentink, D. Flapping wing aerodynamics: From insects to vertebrates. *J. Exp. Biol.* **2016**, *219*, 920–932. [CrossRef] [PubMed]
3. Liu, H.; Ravi, S.; Kolomenskiy, D.; Tanaka, H. Biomechanics and biomimetics in insect-inspired flight systems. *Phil. Trans. R. Soc. Lond. B* **2016**, *371*, 20150390. [CrossRef] [PubMed]
4. Wang, Z.J. Insect flight: From Newton's law to neurons. *Ann. Rev. Condens. Matter Phys.* **2016**, *7*, 281–300. [CrossRef]
5. Shyy, W.; Kang, C.-k.; Chirarattananon, P.; Ravi, S.; Liu, H. Aerodynamics, sensing and control of insect-scale flapping-wing flight. *Proc. Roy. Soc. Lond. A* **2016**, *472*, 20150712. [CrossRef]
6. Shyy, W.; Lian, Y.; Tang, J.; Viieru, D.; Liu, H. *Aerodynamics of Low Reynolds Number Flyers*; Cambridge University Press: Cambridge, UK, 2008.
7. Lehmann, F.-O. The mechanisms of lift enhancement in insect flight. *Naturwissenschaften* **2004**, *91*, 101–122. [CrossRef]
8. Lehmann, F.-O. When wings touch wakes: Understanding locomotor force control by wake–wing interference in insect wings. *J. Exp. Biol.* **2008**, *211*, 224–233. [CrossRef]
9. Sane, S. The aerodynamics of insect flight. *J. Exp. Biol.* **2003**, *206*, 4191–4208. [CrossRef]
10. Wang, Z.J. Dissecting insect flight. *Annu. Rev. Fluid Mech.* **2005**, *37*, 183–210. [CrossRef]
11. Salami, E.; Ward, T.A.; Montazer, E.; Ghazali, N.N.N. A review of aerodynamic studies on dragonfly flight. *J. Mech. Eng. Sci.* **2019**, *233*, 6519–6537. [CrossRef]

12. Vishnudas, V.; Vigoreaux, J.O. Sustained high power performance: Possible strategies for integrating energy supply and demands in flight muscle. In *Nature's Versatile Engine: Insect Flight Muscle Inside and Out*; Vigoreaux, J.O., Ed.; Springer: New York, NY, USA, 2006; p. 288. [CrossRef]

13. Lehmann, F.-O. Muscle Systems design and integration. In *Nature's Versatile Engine: Insect Flight Muscle Inside and Out*; Vigoreaux, J.O., Ed.; Landes Bioscience: Georgetown, DC, USA, 2006; pp. 230–239.

14. Ellington, C.P. Limitations on animal flight performance. *J. Exp. Biol.* **1991**, *160*, 71–91.

15. Lehmann, F.-O.; Bartussek, J. Neural control and precision of flight muscle activation in *Drosophila*. *J. Comp. Physiol. A* **2016**, 1–14. [CrossRef] [PubMed]

16. Sane, S.P. Neurobiology and biomechanics of flight in miniature insects. *Curr. Opin. Neurobiol.* **2016**, *41*, 158–166. [CrossRef]

17. Lehmann, F.-O. Neural control and precision of spike phasing in flight muscles. *J. Neurol. Neuromedicine* **2017**, *2*, 15–19. [CrossRef]

18. Bomphrey, R.J.; Godoy-Diana, R. Insect and insect-inspired aerodynamics: Unsteadiness, structural mechanics and flight control. *Curr. Opin. Insect. Sci.* **2018**, *30*, 26–32. [CrossRef]

19. Sun, X.; Gong, X.; Huang, D. A review on studies of the aerodynamics of different types of maneuvers in dragonflies. *Arch. Appl. Mech.* **2017**, *87*, 521–554. [CrossRef]

20. Taylor, G.K. Mechanics and aerodynamics of insect flight control. *Biol. Rev.* **2001**, *76*, 449–471. [CrossRef]

21. Dickinson, M.H.; Muijres, F.T. The aerodynamics and control of free flight manoeuvres in *Drosophila*. *Phil. Trans. R. Soc. Lond. B* **2016**, *371*, 20150388. [CrossRef]

22. Ellington, C.P. Power and efficiency of insect flight muscle. *J. Exp. Biol.* **1985**, *115*, 293–304.

23. Lehmann, F.-O. The efficiency of aerodynamic force production in *Drosophila*. *Comp. Biochem. Physiol. A* **2001**, *131*, 77–88. [CrossRef]

24. Vincent, J.F.V.; Wegst, U.G.K. Design and mechanical properties of insect cuticle. *Arthropod. Struct. Dev.* **2004**, *33*, 187–199. [CrossRef] [PubMed]

25. Song, F.; Xiao, K.W.; Bai, K.; Bai, Y.L. Microstructure and nanomechanical properties of the wing membrane of dragonfly. *Mat. Sci. Eng. A* **2007**, *457*, 254–260. [CrossRef]

26. Wang, X.-S.; Li, Y.; Shi, Y.-F. Effects of sandwich microstructures on mechanical behaviors of dragonfly wing vein. *Comp. Sci. Tech.* **2008**, *68*, 186–192. [CrossRef]

27. Ma, Y.; Ren, H.; Ning, J.; Zhang, P. Functional morphology and bending characteristics of the honeybee forewing. *J. Bionic. Eng.* **2017**, *14*, 111–118. [CrossRef]

28. Rees, C.J.C. Form and function in corrugated insect wings. *Nature* **1975**, *256*, 200–203. [CrossRef]

29. Combes, S.A.; Daniel, T.L. Flexural stiffness in insect wings I. Scaling and the influence of wing venation. *J. Exp. Biol.* **2003**, *206*, 2979–2987. [CrossRef]

30. Ennos, A.R. Comparative functional morpholoy of the wings of Diptera. *Zool. J. Linn. Soc.* **1989**, *96*, 27–47. [CrossRef]

31. Rajabi, H.; Rezasefat, M.; Darvizeh, A.; Dirks, J.H.; Eshghi, S.; Shafiei, A.; Mostofi, T.M.; Gorb, S.N. A comparative study of the effects of constructional elements on the mechanical behaviour of dragonfly wings. *Appl. Physics A* **2016**, *122*, 19. [CrossRef]

32. Rajabi, H.; Shafiei, A.; Darvizeh, A.; Dirks, J.-H.; Appel, E.; Gorb, S.N. Effect of microstructure on the mechanical and damping behaviour of dragonfly wing veins. *R. Soc. Open Sci.* **2016**, *3*, 160006. [CrossRef]

33. Sunada, S.; Zeng, L.; Kawachi, K. The relationship between dragonfly wing structure and torsional deformation. *J. Theor. Biol.* **1998**, *193*, 39–45. [CrossRef]

34. Meyers, M.A.; Chen, P.-Y.; Lin, A.Y.-M.; Seki, Y. Biological materials: Structure and mechanical properties. *Prog. Mat. Sci.* **2008**, *53*, 1–206. [CrossRef]

35. Wootton, R.J. Support and deformability in insect wings. *J. Zoo. Lond.* **1981**, *193*, 447–468. [CrossRef]

36. Wootton, R.J. The mechanical design of insect wings. *Sci. Am.* **1990**, *263*, 114–121. [CrossRef]

37. Combes, S.A.; Daniel, T.L. Flexural stiffness in insect wings II. Spatial distribution and dynamic wing bending. *J. Exp. Biol.* **2003**, *206*, 2989–2997. [CrossRef]

38. Combes, S.A.; Daniel, T.L. Into thin air: Contributions of aerodynamic and intertial-elastic forces to wing bending in the hawkmoth *Manduca sexta*. *J. Exp. Biol.* **2003**, *206*, 2999–3006. [CrossRef]

39. Appel, E.; Heepe, L.; Lin, C.-P.; Gorb, S.N. Ultrastructure of dragonfly wing veins: Composite structure of fibrous material supplemented by resilin. *J. Anat.* **2015**, *227*, 561–582. [CrossRef]

40. Dirks, J.-H.; Taylor, D. Veins improve fracture toughness of insect wings. *PLoS ONE* **2012**, *7*. [CrossRef]

41. Rajabi, H.; Darvizeh, A.; Shafiei, A.; Taylor, D.; Dirks, J.-H. Numerical investigation of insect wing fracture behaviour. *J. Biomech.* **2015**, *48*, 89–94. [CrossRef]

42. Agrawal, S.; Grimaldi, D.; Fox, J.L. Haltere morphology and campaniform sensilla arrangement across Diptera. *Arthropod. Struct. Dev.* **2017**, *46*, 215–229. [CrossRef]

43. Pratt, B.; Deora, T.; Mohren, T.; Daniel, T. Neural evidence supports a dual sensory-motor role for insect wings. *Proc. Roy. Soc. Lond. B* **2017**, *284*, 20170969. [CrossRef]

44. Dickinson, M.H. Comparison of encoding properties of campaniform sensilla on the fly wing. *J. Exp. Biol.* **1990**, *151*, 245–261.

45. Cole, E.S.; Palka, J. The pattern of campaniform sensilla on the wing and haltere of *Drosophila melanogaster* and several of its homeotic mutants. *Development* **1982**, *71*, 41–61.

46. Gnatzy, W.; Grünert, U.; Bender, M. Campaniform sensilla of *Calliphora vicina* (Insecta, Diptera): I. Topography. *Zoomorphology* **1987**, *106*, 312–319. [CrossRef]

47. Ray, R.P.; Nakata, T.; Henningsson, P.; Bomphrey, R.J. Enhanced flight performance by genetic manipulation of wing shape in *Drosophila*. *Nat. Comm.* **2016**, *7*, 1–8. [CrossRef]

48. Gorb, S.; Kesel, A.; Berger, J. Microsculpture of the wing surface in Odonata: Evidence for cuticular wax covering. *Arthropod. Struct. Dev.* **2000**, *29*, 129–135. [CrossRef]

49. Wootton, R.J. Functional morphology of insect wings. *Annu. Rev. Entomol.* **1992**, *37*, 113–140. [CrossRef]

50. Luo, G.; Sun, M. The effects of corrugation and wing planform on the aerodynamic force production of sweeping model insect wings. *Acta Mech. Sin.* **2005**, *21*, 531–541. [CrossRef]

51. Koehler, C.; Liang, Z.; Gaston, Z.; Wan, H.; Dong, H. 3D reconstruction and analysis of wing deformation in free-flying dragonflies. *J. Exp. Biol.* **2012**, *215*, 3018–3027. [CrossRef]

52. Jongerius, S.R.; Lentink, D. Structural analysis of a dragonfly wing. *Exp. Mech.* **2010**, *50*, 1323–1334. [CrossRef]

53. Barnes, C.J.; Visbal, M.R. Numerical exploration of the origin of aerodynamic enhancements in [low-Reynolds number] corrugated airfoils. *Phys. Fluids* **2013**, *25*, 115106. [CrossRef]

54. Feaster, J.; Battaglia, F.; Bayandor, J. A computational study on the influence of insect wing geometry on bee flight mechanics. *Biol. Open* **2017**, *6*, 1784–1795. [CrossRef] [PubMed]

55. Le, T.Q.; Truong, T.V.; Tran, H.T.; Park, S.H.; Ko, J.H.; Park, H.C.; Yoon, K.J.; Byun, D. Two- and three-dimensional simulations of beetle hind wing flapping during free forward flight. *J. Bionic. Eng.* **2013**, *10*, 316–328. [CrossRef]

56. Okamoto, M.; Yasuda, K.; Azuma, A. Aerodynamic characteristics of the wings and body of a dragonfly. *J. Exp. Biol.* **1996**, *199*, 281–294. [PubMed]

57. Harbig, R.R.; Sheridan, J.; Thompson, M.C. Reynolds number and aspect ratio effects on the leading-edge vortex for rotating insect wing planforms. *J. Fluid Mech.* **2013**, *717*, 166–192. [CrossRef]

58. Jain, S.; Bhatt, V.D.; Mittal, S. Shape optimization of corrugated airfoils. *Comp. Mech.* **2015**, *56*, 917–930. [CrossRef]

59. Meng, X.G.; Mao, S. Aerodynamic effects of wing corrugation at gliding flight at low Reynolds numbers. *Phys. Fluids* **2013**, *25*, 071905. [CrossRef]

60. Brandt, J.; Doig, G.; Tsafnat, N. Computational aerodynamic analysis of a micro-CT based bio-realistic fruit fly wing. *PLoS ONE* **2015**, *10*, e0124824. [CrossRef]

61. Rees, C.J.C. Aerodynamic properties of an insect wing section and a smooth aerofoil compared. *Nature* **1975**, *258*, 141–142. [CrossRef]

62. Ortega Ancel, A.; Eastwood, R.; Vogt, D.; Ithier, C.; Smith, M.; Wood, R.; Kovač, M. Aerodynamic evaluation of wing shape and wing orientation in four butterfly species using numerical simulations and a low-speed wind tunnel, and its implications for the design of flying micro-robots. *Interface focus* **2017**, *7*, 20160087. [CrossRef]

63. New, T.H.; Chan, Y.X.; Koh, G.C.; Hoang, M.C.; Shi, S. Effects of corrugated aerofoil surface features on flow-separation control. *AIAA J.* **2014**, *52*, 206–211. [CrossRef]

64. Murphy, J.T.; Hu, H. An experimental study of a bio-inspired corrugated airfoil for micro air vehicle applications. *Exp. Fluids* **2010**, *49*, 531–546. [CrossRef]

65. Chen, Y.; Skote, M. Gliding performance of 3-D corrugated dragonfly wing with spanwise variation. *J. Fluid Struct.* **2016**, *62*, 1–13. [CrossRef]

66. Ansari, M.I.; Anwer, S.F. Numerical analysis of an insect wing in gliding flight: Effect of corrugation on suction side. *FDMP* **2018**, *14*, 259–279. [CrossRef]

67. Reiser, M.B.; Dickinson, M.H. A test bed for insect-inspired robotic control. *Phil. Trans. R. Soc. Lond. A* **2003**, *361*, 2267–2285. [CrossRef]

68. Wang, Z.J.; Birch, J.M.; Dickinson, M.H. Unsteady forces and flows in low Reynolds number hovering flight: Two-dimensional computations *vs* robotic wing experiments. *J. Exp. Biol.* **2004**, *207*, 449–460. [CrossRef] [PubMed]

69. Fry, S.N.; Sayaman, R.; Dickinson, M.H. The aerodynamics of hovering flight in *Drosophila*. *J. Exp. Biol.* **2005**, *208*, 2303–2318. [CrossRef]

70. Sane, S.; Dickinson, M.H. The control of flight force by a flapping wing: Lift and drag production. *J. Exp. Biol.* **2001**, *204*, 2607–2626.

71. Birch, J.M.; Dickinson, M.H. The influence of wing-wake interactions on the production of aerodynamic forces in flapping flight. *J. Exp. Biol.* **2003**, *206*, 2257–2272. [CrossRef]

72. Dickinson, M.H.; Lehmann, F.-O.; Sane, S. Wing rotation and the aerodynamic basis of insect flight. *Science* **1999**, *284*, 1954–1960. [CrossRef]

73. Lehmann, F.-O. Wing–wake interaction reduces power consumption in insect tandem wings. *Exp. Fluids* **2008**, *46*, 765–775. [CrossRef]

74. Maybury, W.J.; Lehmann, F.-O. The fluid dynamics of flight control by kinematic phase lag variation between two robotic insect wings. *J. Exp. Biol.* **2004**, *207*, 4707–4726. [CrossRef] [PubMed]

75. Wehmann, H.-N.; Heepe, L.; Gorb, S.N.; Engels, T.; Lehmann, F.-O. Local deformation and stiffness distribution in fly wings. *Biol. Open* **2019**, *8*, bio038299. [CrossRef] [PubMed]

76. Shyy, W.; Aono, H.; Chimakurthi, S.K.; Trizila, P.; Kang, C.-K.; Cesnik, C.E.S.; Liu, H. Recent progress in flapping wing aerodynamics and aeroelasticity. *Prof. Aero. Sci.* **2010**, *46*, 284–327. [CrossRef]

77. Bluman, J.E.; Sridhar, M.K.; Kang, C.-k. Chordwise wing flexibility may passively stabilize hovering insects. *J. R. Soc. Interface* **2018**, *15*, 20180409. [CrossRef]

78. Moses, K.C.; Michaels, S.C.; Willis, M.; Quinn, R.D. Artificial *Manduca sexta* forewings for flapping-wing micro aerial vehicles: How wing structure affects performance. *Bioinsp. Biomim.* **2017**, *12*, 055003. [CrossRef]

79. Mountcastle, A.M.; Daniel, T.L. Aerodynamic and functional consequences of wing compliance. *Exp. Fluids* **2009**, *46*, 873–882. [CrossRef]

80. Nakata, T.; Liu, H. Aerodynamic performance of a hovering hawkmoth with flexible wings: A computational approach. *Proc. Roy. Soc. Lond. B* **2012**, *279*, 722–731. [CrossRef]

81. Tanaka, H.; Whitney, J.P.; Wood, R.J. Effect of flexural and torsional wing flexibility on lift generation in hoverfly flight. *Integ. Comp. Biol.* **2011**, *51*, 142–150. [CrossRef] [PubMed]

82. Tobing, S.; Young, J.; Lai, J.C.S. Effects of wing flexibility on bumblebee propulsion. *J. Fluid Struct.* **2017**, *68*, 141–157. [CrossRef]

83. Engels, T.; Wehmann, H.-N.; Lehmann, F.-O. Three-dimensional wing structure attenuates aerodynamic efficiency in flapping fly wings. *J. R. Soc. Interface* **2020**, *17*, 20190804. [CrossRef]

84. Stevens, R.J.; Babinsky, H.; Manar, F.; Mancini, P.; Jones, A.R.; Granlund, K.O.; Ol, M.V.; Nakata, T.; Phillips, N.; Bomphrey, R. Low Reynolds number acceleration of flat plate wings at high incidence. In Proceedings of the 54th AIAA Aerospace Sciences Meeting, San Diego, CA, USA, 4–8 January 2016; p. 0286. [CrossRef]

85. Usherwood, J.R.; Ellington, C.P. The aerodynamics of revolving wings II. propeller force coefficients from mayfly to quail. *J. Exp. Biol.* **2002**, *205*, 1565–1576. [PubMed]

86. Usherwood, J.R.; Ellington, C.P. The aerodynamic of revolving wings I. model hawkmoth wings. *J. Exp. Biol.* **2002**, *205*, 1547–1564. [PubMed]

87. Jones, A.R.; Manar, F.; Phillips, N.; Nakata, T.; Bomphrey, R.; Ringuette, M.J.; Percin, M.; van Oudheusden, B.; Palmer, J. Leading edge vortex evolution and lift production on rotating wings. In Proceedings of the 54th AIAA Aerospace Sciences Meeting, San Diego, CA, USA, 4–8 January 2016; p. 0288. [CrossRef]

88. Rival, D.; Tropea, C. Characteristics of pitching and plunging airfoils under dynamic-stall conditions. *J. Aircraft* **2010**, *47*, 80–86. [CrossRef]

89. Jantzen, R.T.; Taira, K.; Granlund, K.O.; Ol, M.V. Vortex dynamics around pitching plates. *Phys. Fluids* **2014**, *26*, 053606. [CrossRef]

90. Manar, F.; Medina, A.; Jones, A.R. Tip vortex structure and aerodynamic loading on rotating wings in confined spaces. *Exp. Fluids* **2014**, *55*, 1815. [CrossRef]

91. Lua, K.B.; Zhang, X.; Lim, T.; Yeo, K. Effects of pitching phase angle and amplitude on a two-dimensional flapping wing in hovering mode. *Exp. Fluids* **2015**, *56*, 35. [CrossRef]
92. Ellington, C.P.; Berg, C.v.d.; Willmott, A.P.; Thomas, A.L.R. Leading-edge vortices in insect flight. *Nature* **1996**, *384*, 626–630. [CrossRef]
93. Lehmann, F.O.; Sane, S.P.; Dickinson, M.H. The aerodynamic effects of wing-wing interaction in flapping insect wings. *J. Exp. Biol.* **2005**, *208*, 3075–3092. [CrossRef]
94. Bomphrey, R.J.; Nakata, T.; Phillips, N.; Walker, S.M. Smart wing rotation and trailing-edge vortices enable high frequency mosquito flight. *Nature* **2017**, *544*, 92–95. [CrossRef]
95. Cheng, B.; Sane, S.P.; Barbera, G.; Troolin, D.R.; Strand, T.; Deng, X. Three-dimensional flow visualization and vorticity dynamics in revolving wings. *Exp. Fluids* **2013**, *54*, 1423. [CrossRef]
96. Garmann, D.; Visbal, M. Dynamics of revolving wings for various aspect ratios. *J. Fluid Mech.* **2014**, *748*, 932–956. [CrossRef]
97. Percin, M.; Van Oudheusden, B. Three-dimensional flow structures and unsteady forces on pitching and surging revolving flat plates. *Exp. Fluids* **2015**, *56*, 47. [CrossRef]
98. Wolfinger, M.; Rockwell, D. Transformation of flow structure on a rotating wing due to variation of radius of gyration. *Exp. Fluids* **2015**, *56*, 137. [CrossRef]
99. Carr, Z.R.; DeVoria, A.C.; Ringuette, M.J. Aspect-ratio effects on rotating wings: Circulation and forces. *J. Fluid Mech.* **2015**, *767*, 497–525. [CrossRef]
100. Krishna, S.; Green, M.A.; Mulleners, K. Effect of pitch on the flow behavior around a hovering wing. *Exp. Fluids* **2019**, *60*, 86. [CrossRef]
101. Ellington, C.P. The novel aerodynamics of insect flight: Applications to micro-air vehicles. *J. Exp. Biol.* **1999**, *202*, 3439–3448.
102. Pick, S.; Lehmann, F.-O. Stereoscopic PIV on multiple color-coded light sheets and its application to axial flow in flapping robotic insect wings. *Exp. Fluids* **2009**, *47*, 1009–1023. [CrossRef]
103. Birch, J.M.; Dickinson, M.H. Spanwise flow and the attachment of the leading-edge vortex on insect wings. *Nature* **2001**, *412*, 729–733. [CrossRef]
104. Ellington, C.P. The aerodynamics of hovering insect flight. V. A vortex theory. *Phil. Trans. R. Soc. Lond. B* **1984**, *305*, 115–144.
105. Muijres, F.T.; Spedding, G.R.; Winter, Y.; Hedenström, A. Actuator disk model and span efficiency of flapping flight in bats based on time-resolved PIV measurements. *Exp. Fluids* **2011**, *51*, 511–525. [CrossRef]
106. Milne-Thomson, L.M. *Theoretical Aerodynamics*, 4th ed.; Macmillan: New York, NY, USA, 1966; pp. 210–211.
107. Spedding, G.; McArthur, J. Span efficiencies of wings at low Reynolds numbers. *J. Aircraft* **2010**, *47*, 120–128. [CrossRef]
108. Henningsson, P.; Bomphrey, R.J. Span efficiency in hawkmoths. *J. R. Soc. Interface* **2013**, *10*, 20130099. [CrossRef]
109. Prandtl, L. Tragflügeltheorie. I. Mitteilung. *Nachricht. Gesell. Wissensch. Göttingen* **1918**, *1918*, 451–477.
110. Betz, A. Schraubenpropeller mit geringstem Energieverlust. *Nachricht. Gesell. Wissensch. Göttingen* **1919**, *1919*, 193–217.
111. Goldstein, S. On the vortex theory of screw propellers. *Proc. Roy. Soc. Lond. A* **1929**, *123*, 440–465.
112. Larrabee, E.E.; French, S.E. Minimum induced loss windmills and propellers. *J. Wind Eng. Industr. Aero.* **1983**, *15*, 317–327. [CrossRef]
113. Nabawy, M.R.; Crowther, W.J. Is Flapping Flight Aerodynamically Efficient? In Proceedings of the 32nd AIAA Applied Aerodynamics Conference, Atlanta, GA, USA, 16–20 June 2014; p. 2277.
114. Nabawy, M.R.; Crowther, W.J. Aero-optimum hovering kinematics. *Bioinsp. Biomim.* **2015**, *10*, 044002. [CrossRef]
115. Nabawy, M.R.; Crowther, W.J. Optimum hovering wing planform. *J. Theor. Biol.* **2016**, *406*, 187–191. [CrossRef]
116. Reid, H.E.; Schwab, R.K.; Maxcer, M.; Peterson, R.K.D.; Johnson, R.L.; Jankauski, M. Wing flexibility reduces the energetic requirements of insect flight. *Bioinsp. Biomim.* **2019**, *14*, 056007. [CrossRef]
117. Bhat, S.; Zhao, J.; Sheridan, J.; Hourigan, K.; Thompson, M. Aspect ratio studies on insect wings. *Phys. Fluids* **2019**, *31*, 121301. [CrossRef]
118. Levy, D.-E.; Seifert, A. Simplified dragonfly airfoil aerodynamics at Reynolds numbers below 8000. *Phys. Fluids* **2009**, *21*, 071901. [CrossRef]

119. Weis-Fogh, T. Quick estimates of flight fitness in hovering animals, including novel mechanisms for lift production. *J. Exp. Biol.* **1973**, *59*, 169–230.

120. Ellington, C.P. The aerodynamics of hovering insect flight. II. Morphological parameters. *Phil. Trans. R. Soc. Lond. B* **1984**, *305*, 17–40.

121. Ennos, A.R. The effect of size on the optimal shapes of gliding insects and seeds. *J. Zoo. Lond.* **1989**, *219*, 61–69. [CrossRef]

122. Zanker, J.M. The wing beat of *Drosophila melanogaster* I. Kinematics. *Phil. Trans. R. Soc. Lond. B* **1990**, *327*, 1–18.

123. Han, J.-S.; Chang, J.W.; Cho, H.-K. Vortices behavior depending on the aspect ratio of an insect-like flapping wing in hover. *Exp. Fluids* **2015**, *56*, 181. [CrossRef]

124. Suzuki, K.; Yoshino, M. A trapezoidal wing equivalent to a Janatella leucodesma's wing in terms of aerodynamic performance in the flapping flight of a butterfly model. *Bioinsp. Biomim.* **2019**, *14*, 036003. [CrossRef]

125. Engels, T.; Kolomenskiy, D.; Schneider, K.; Sesterhenn, J. FluSI: A novel parallel simulation tool for flapping insect flight using a Fourier method with volume penalization. *SIAM J. Sci. Comp.* **2016**, *38*, S3–S24. [CrossRef]

126. Sroka, M.; Engels, T.; Krah, P.; Mutzel, S.; Schneider, K.; Reiss, J. An open and parallel multiresolution framework using block-based adaptive grids. In *Active Flow and Combustion Control 2018*; Springer: Berlin/Heidelberg, Germany, 2019; pp. 305–319.

127. Usherwood, J.R.; Lehmann, F.-O. Phasing of dragonfly wings can improve aerodynamic efficiency by removing swirl. *J. R. Soc. Interface* **2008**, *5*, 1303–1307. [CrossRef]

128. Ellington, C.P. The aerodynamics of hovering insect flight. VI. Lift and power requirements. *Phil. Trans. R. Soc. Lond. B* **1984**, *305*, 145–181.

129. Phillips, N.; Knowles, K.; Lawson, N. Effect of wing planform shape on the flow structures of an insect-like flapping wing in hover. In Proceedings of the 27th International Congress of the Aeronautical Sciences ICAS, Nice, France, 19–24 September 2010.

130. Lu, Y.; Shen, G.X.; Lai, G.J. Dual leading-edge vortices on flapping wings. *J. Exp. Biol.* **2006**, *209*, 5005–5016. [CrossRef] [PubMed]

131. Rival, D.E.; Kriegseis, J.; Schaub, P.; Widmann, A.; Tropea, C. Characteristic length scales for vortex detachment on plunging profiles with varying leading-edge geometry. *Exp. Fluids* **2014**, *55*, 1660. [CrossRef]

132. Wootton, R.J. Leading edge section and asymmetric twisting in the wings of flying butterflies (insecta, papilionoidea). *J. Exp. Biol.* **1993**, *180*, 105–117.

133. Newman, D.J.S.; Wooton, R.J. An approach to the mechanics of pleating in dragonfly wings. *J. Exp. Biol.* **1986**, *125*, 361–372.

134. Lian, Y.; Broering, T.; Hord, K.; Prater, R. The characterization of tandem and corrugated wings. *Progr. Aero. Sci.* **2014**, *65*, 41–69. [CrossRef]

135. Kesel, A.B. Aerodynamic characteristics of dragonfly wing sections compared with technical aerofoils. *J. Exp. Biol.* **2000**, *203*, 3125–3135.

136. Kim, W.-K.; Ko, J.H.; Park, H.C.; Byun, D. Effects of corrugation of the dragonfly wing on gliding performance. *J. Theor. Biol.* **2009**, *260*, 523–530. [CrossRef]

137. Huda, N.; Anwer, S.F. *The Effects of Leading Edge Orientation on the Aerodynamic Performance of Dragon Fly Wing Section in Gliding Flight*; Springer: New Delhi, India, 2016; pp. 1433–1441.

138. Hord, K.; Liang, Y. Numerical investigation of the aerodynamic and structural characteristics of a corrugated airfoil. *J. Aircraft* **2012**, *49*, 749–757. [CrossRef]

139. Xiang, J.; Du, J.; Li, D.; Liu, K. Aerodynamic performance of the locust wing in gliding mode at low Reynolds number. *J. Bionic. Eng.* **2016**, *13*, 249–260. [CrossRef]

140. Shahzad, A.; Hamdani, H.R.; Aizaz, A. Investigation of corrugated wing in unsteady motion. *J. Appl. Fluid Dyn.* **2017**, *10*, 833–845. [CrossRef]

141. Meng, X.; Sun, M. Aerodynamic effects of corrugation in flapping insect wings in forward flight. *J. Bionic. Eng.* **2011**, *8*, 140–150. [CrossRef]

142. Du, G.; Sun, M. Aerodynamic effects of corrugation and deformation in flapping wings of hovering hoverflies. *J. Theor. Biol.* **2012**, *300*, 19–28. [CrossRef] [PubMed]

143. Flint, T.J.; Jermy, M.C.; New, T.H.; Ho, W.H. Computational study of a pitching bio-inspired corrugated airfoil. *Int. J. Heat Fluid Flow* **2017**, *65*, 328–341. [CrossRef]

144. Vargas, A.; Mittal, R.; Dong, H. A computational study of the aerodynamic performance of a dragonfly wing section in gliding flight. *Bioinsp. Biomim.* **2008**, *3*, 026004. [CrossRef]

145. Du, G.; Sun, M. Effects of wing deformation on aerodynamic forces in hovering hoverflies. *J. Exp. Biol.* **2010**, *213*, 2273–2283. [CrossRef]

146. Haas, F.; Gorb, S.N.; Wootton, R.J. Elastic joints in dermapteran hind wings: Materials and wing folding. *Arthropod. Struct. Dev.* **2000**, *29*, 137–146. [CrossRef]

147. Hou, D.; Zhong, Z.; Yin, Y.; Pan, Y.; Zhao, H. The role of soft vein joints in dragonfly flight. *J. Bionic. Eng.* **2017**, *14*, 738–745. [CrossRef]

148. Rajabi, H.; Ghoroubi, N.; Darvizeh, A.; Dirks, J.-H.; Appel, E.; Gorb, S.N. A comparative study of the effects of vein-joints on the mechanical behaviour of insect wings: I. Single joints. *Bioinsp. Biomim.* **2015**, *10*, 056003. [CrossRef]

149. Andersen, S.O.; Weis-Fogh, T. Resilin, a rubber-like protein in arthropod cuticle. *Adv. Insect Physiol.* **1964**, *2*, 1–65.

150. Appel, E.; Gorb, S.N. Resilin-bearing wing vein joints in the dragonfly *Epiophlebia superstes*. *Bioinsp. Biomim.* **2011**, *6*. [CrossRef]

151. Gorb, S.N. Serial elastic elements in the damselfly wing: Mobile vein joints contain resilin. *Naturwissenschaften* **1999**, *86*, 552–555. [CrossRef]

152. Kovalev, A.; Filippov, A.; Gorb, S.N. Slow viscoelastic response of resilin. *J. Comp. Physiol. A* **2018**, *204*, 409–417. [CrossRef] [PubMed]

153. Qin, G.; Hu, X.; Cebe, P.; Kaplan, D.L. Mechanism of resilin elasticity. *Nat. Comm.* **2012**, *3*, 1–9. [CrossRef] [PubMed]

154. Wong, D.C.; Pearson, R.D.; Elvin, C.M.; Merritt, D.J. Expression of the rubber-like protein, resilin, in developing and functional insect cuticle determined using a *Drosophila* anti-rec 1 resilin antibody. *Dev. Dyn.* **2012**, *241*, 333–339. [CrossRef] [PubMed]

155. Weis-Fogh, T. A rubber-like protein in insect cuticles. *J. Exp. Biol.* **1960**, *37*, 889–907.

156. Jensen, M.; Weis-Fogh, T. Biology and physics of locust flight. V. Strength and elasticity of locust cuticle. *Phil. Trans. Roy. Soc. London B* **1962**, *245*, 137–169.

157. Biewener, A.A.; Roberts, T.J. Muscle and tendon contributions to force, work, and elastic energy savings: A comparative perspective. *Exerc. Sport Sci. Rev.* **2000**, *28*, 99–107.

158. Lichtwark, G.A.; Barclay, C.J. The influence of tendon compliance on muscle power output and efficiency during cyclic contractions. *J. Exp. Biol.* **2010**, *213*, 707–714. [CrossRef]

159. Roberts, T.J.; Marsh, R.L.; Weyand, P.G.; Taylor, C.R. Muscular force in running turkeys: The economy of minimizing work. *Science* **1997**, *275*, 1113–1115. [CrossRef]

160. Zhao, L.; Huang, Q.; Deng, X.; Sane, S.P. Aerodynamic effects of flexibility in flapping wings. *J. R. Soc. Interface* **2010**, *7*, 485–497. [CrossRef]

161. Lehmann, F.-O.; Gorb, S.; Nasir, N.; Schützner, P. Elastic deformation and energy loss of flapping fly wings. *J. Exp. Biol.* **2011**, *l214*, 2949–2961. [CrossRef] [PubMed]

162. Wootton, R.J. Invertebrate paraxial locomotory appendages: Design, deformation and control. *J. Exp. Biol.* **1999**, *202*, 3333–3345. [PubMed]

163. Young, J.; Walker, S.M.; Bomphrey, R.J.; Taylor, G.K.; Thomas, A.L.R. Details of insect wing design and deformation enhance aerodynamic function and flight efficiency. *Science* **2009**, *325*, 1549–1552. [CrossRef] [PubMed]

164. Zheng, L.; Hedrick, T.L.; Mittal, R. Time-varying wing-twist improves aerodynamic efficiency of forward flight in butterflies. *PLoS ONE* **2013**, *8*, e53060. [CrossRef]

165. Mistick, E.A.; Mountcastle, A.M.; Combes, S.A. Wing flexibility improves bumblebee flight stability. *J. Exp. Biol.* **2016**, *219*, 3384–3390. [CrossRef]

166. Nguyen, T.T.; Sundar, D.S.; Yeo, K.S.; Lim, T.T. Modeling and analysis of insect-like flexible wings at low Reynolds number. *J. Fluid Struct.* **2016**, *62*, 294–317. [CrossRef]

 *insects*

Article

Comparison of Wing, Ovipositor, and Cornus Morphologies between *Sirex noctilio* and *Sirex nitobei* Using Geometric Morphometrics

Ming Wang [1,2], Lixiang Wang [3], Ningning Fu [1,2], Chenglong Gao [1,2], Tegen Ao [4], Lili Ren [1,2,*] and Youqing Luo [1,2,*]

[1] Beijing Key Laboratory for Forest Pest Control, Beijing Forestry University, Beijing 100083, China; 13020028768@163.com (M.W.); funingning2012@sina.com (N.F.); gaocl890907@163.com (C.G.)
[2] Sino-French joint Laboratory for Invasive Forest Pests in Eurasia, INRA-Beijing Forestry University, Beijing 100083, China
[3] College of Plant Protection, Gansu Agricultural University, Lanzhou 730070, China; wanglx@gsau.edu.cn
[4] Tongliao Control and Quarantine Station of Forest Pest, Tongliao 028000, China; autogen1@163.com
* Correspondence: lily_ren@bjfu.edu.cn (L.R.); yqluo@bjfu.edu.cn (Y.L.)

Received: 31 December 2019; Accepted: 21 January 2020; Published: 24 January 2020

Abstract: *Sirex noctilio* F. (Hymenoptera: Siricidae) is an invasive woodwasp from Europe and North Africa. Globalization has led to an expanding global presence in pine forests. *S. noctilio* has been previously introduced outside of its native range and now co-occurs in trees with native *S. nitobei* Matsumura (first discovered in 2016). Damage to *Pinus sylvestris* var. *mongolica* Litv in northeast China can be attributed to two types of woodwasp. To distinguish the two species by the traditional taxonomic morphology, we mainly differentiate the color of the male's abdomen and the female's leg. There remains intraspecific variation like leg color in the delimitation of related genera or sibling species of *Sirex* woodwasps. In this study, we used landmark-based geometric morphometrics including principal component analysis, canonical variate analysis, thin-plate splines, and cluster analysis to analyze and compare the wings, ovipositors, and cornus of two woodwasps to ascertain whether this approach is reliable for taxonomic studies of this group. The results showed significant differences in forewing venation and the shapes of pits in the middle of ovipositors among the two species, whereas little difference in hindwings and cornus was observed. This study assists in clarifying the taxonomic uncertainties of Siricidae and lays a foundation for further studies of the interspecific relationships of the genus *Sirex*.

Keywords: cornus; geometric morphometrics; ovipositor; Siricidae; taxonomy; wing

1. Introduction

Siricidae is a small group of species, in which individuals are relatively large with a clean body surface and easily identifiable morphological features. One of the most striking features of Siricidae is what appears to be incredible variation in wing venation, including the appearance or the disappearance of veins symmetrically or asymmetrically on either wing. Such variation is very rarely seen in other Hymenoptera, a group where wing veins are important for classification [1]. The wing characteristics of Siricidae are unstable and seldom used in taxonomic studies. In general, the classification and identification of this group is based mainly on the structure of the thorax and abdomen. However, these structures are very similar in closely related species; thus, it is difficult to accurately identify species in some cases. Interestingly, geometrics, as a new classification method, has been recently applied in many classification studies of Hymenoptera [2].

It is well known that insect wing shape can exhibit a high heritability in nature [3,4]. Thus, wing morphology is of primary importance to entomologists studying systematics. Since the 1970s, several investigators have used the two-dimensional characteristics of insect wings to advance the fields of systematics and phylogeny [5,6]. Geometric morphometrics utilizes powerful and comprehensive statistical procedures to analyze the variations in shape using either homologous landmarks or outlines of the structure [6–8], and it is currently considered to be the most rigorous morphometric method. Wings are excellent in studies that define morphological variations because they are nearly two-dimensional and their venation provides many well-defined morphological landmarks [9]. For instance, vein intersections are easily identifiable, which enables the general shape of the wing to be captured [10]. The use of geometric morphometrics to study wing venation has also been useful in insect identification at the individual level [11], in differentiation between sibling species [12,13], and in delimitation among genera. However, thismethodology has not yet been applied in studies of woodwasps.

The wood-boring wasp, *Sirex noctilio* Fabricius (Hymenoptera: Siricidae), is an invasive pest of numerous species of pine tree (*Pinus* spp.) worldwide and most of the destruction from *S. noctilio* is in commercial plants [14,15]. In August 2013, woodwasps were detected as a pest of Mongolian pine (*Pinus sylvestris* var. *mongolica*) in the Duerbote Mongolian Autonomous County, Heilongjiang Province, China [16]. In 2016, two morphologically similar woodwasps were found to damage *P. sylvestris* var. *mongolica* in Jinbaotun forest farm, Inner Mongolia, causing a lot of pine forests to weaken and die. After morphological comparison and molecular identification, the two woodwasps were *S. noctilio* and *S. nitobei* [17]. Each have two pairs of large, transparent-film wings with visible mesh veins. As the insect wing is a planar structure, it is relatively easy to acquire two-dimensional images, and it is difficult to unintentionally distort the structure. Unfortunately, it is rather difficult to identify *S. noctilio* and *S. nitobei* wing vein characteristics with the naked eye, and these two species display unstable vein patterns, which means that the use of geometric morphometrics is appropriate [18].

In previous studies, the pits on the ventral portion (lancet) of the ovipositor have been consistently used as an identifying structure. The lancets of the ovipositor independently slide back and forth to move the egg and to penetrate wood. This characteristic was used for the first time by Kjellander (1945) to segregate females of *S. juvencus* from those of *S. noctilio*. Furthermore, the size, shape, and number of pits on the ovipositors can be used as distinguishing features for the identification of most species [15]. This also holds true for *Sirex*, in which the most important distinguishing characteristics on the ovipositor are pits located from the base to approximately the middle of the lancet, although the apical teeth segments usually do not show distinct differences [15]. Another striking diagnostic feature is the large hornlike projection, called the cornus, on the last abdominal segment of the females. The cornus is thought to help the larvae pack the frass in the tunnel. The cornus varies in shape (the shape of the female cornus does not vary with size for most species), although their distinguishing features remain poorly characterized. These difficulties underline the need for further studies to clarify the taxonomy of woodwasps, either by searching for new morphological characteristics with clear distinguishing variations or applying alternative effective methods to provide a basis for studying flight and reproductive behavior.

Geometric morphometrics [6,19,20] overcomes the shortcomings of conventional morphological analysis and focuses on the topological information of the organic form [18]. In addition, as it is not affected by various factors, such as size and shape, this method has the potential to be more widely used in the identification of insects, resulting in automatic insect recognition system that is continually updated and improved [3]. In taxonomy and other fields, genetics and morphometrics are complementary tools that are often used to understand the origins of phenotypic differences. The application of marker points in biology can be divided into three categories [21]. In this study, we focused on two of these categories, namely (i) the common points that can be accurately found on each specimen based on the anatomical features, which is a mathematical point supported by substantial evidence between homologous subjects [22], such as structural intersections (the basis for marking points on wings); and (ii) the mathematical point for homologous subjects, which is supported by geometrical rather than histological evidence, such as depressed or convex points (the basis for marking points on ovipositors and cornus). Platts analysis was used to superimpose the marker points and minimize the deviation of the marker points. In the same coordinate system, the influence of non-morphological factors in morphological information analysis was eliminated, and the average contour of each population was obtained. Thus, in the present study, we applied landmark-based geometric morphometrics to quantify and analyze wing, cornus, and ovipositor morphologies of two *Sirex* species that have not been previously characterized. We explored the similarities between these species to strengthen the available quantitative research data that form the basis of species identification and to provide new insights for automatic insect identification systems.

2. Materials and Methods

2.1. Ethics Statement

This study did not involve endangered or protected species. No specific permits were required for this study.

2.2. Insects

Insect samples were collected from the Jinbaotun Forest in Tongliao City, Inner Mongolia, from June 2016 to September 2017. Trees were felled in early summer, and insects were collected at emergence. Insects were collected from different, unrelated plots. Sirex specimens were structurally analyzed (Table 1). Prior to geometric morphometric analysis, specimens were identified using adult morphological characteristics, including the color of their thoracic legs, abdomen [1,23].

Table 1. The *Sirex* specimens collection information in this study.

Species	Family	Host	Number		Collecting Date
			Female	Male	
Sirex noctilio	Siricidae	*Pinus sylvestris* var. *mongolica* Litv	20	20	2016/07/15
Sirex nitobei	Siricidae	*Pinus sylvestris* var. *mongolica* Litv	20	20	2016/09/05

2.3. Insect Processing and Image Acquisition

The front and rear wings of each specimen were cut off from the body. Rohlf et al. suggested using only one side of each paired organ or limb to avoid asymmetry bias between the two halves [24]. In this study, only the left wings of specimens were used, which were ultrasonically cleaned with 75% alcohol for 90 s to remove impurities. Thereafter, specimens were dehydrated with an ascending series of ethanol washes (75%, 80%, 85%, 90%, 95%, 100%, and 100%) for 20 min each. Specimens were softened with xylene, placed on glass slides that were previously wiped clean with a lens tissue, and then mounted in neutral balsam.

Ovipositors and cornus were dissected from the *Sirex* specimens, and the remaining parts were stored at −20 °C. We used ten individuals of each species to examine the pits and cornus. Impurities were removed from the specimens by ultrasonic cleaning (Skymen, JP-1200, Shenzhen, China) or brushing. The specimens were then placed face up on clean glass slides and mounted in neutral gum (Coolaber, Beijing, China). All specimens were numbered.

A light microscopy (Leica, S4E, Wetzlar, Germany) was performed to determine the number and distribution of pits on ovipositors. Images were captured with a Nikon camera (Nikon D90, Tokyo, Japan). Each image was saved as a 24-bit. bmp image, and original stored images were used in subsequent analysis rather than compressed files. The directions and positions of the specimen images were readjusted with Photoshop CC2015 software (Adobe Systems, San Jose, USA).

2.4. Standardization of Data and Statistical Analysis

The TPS files were maked from selected images of wings, ovipositors, and cornus using TpsUtil software (tpsUtil 1.47, [24]). The landmarks in each image were recorded as the central location point of each specimen and digitized using TpsDig2 software (New York, NY, USA) [24]. For each. Tps file, the landmarks were scanned in the same order, and the scale factor was set for each image. Therefore, 20 landmarks from the forewing, 11 landmarks from the hindwing (Figure 1a), nine landmarks (Figure 2) from the ovipositor (from the base to the middle pits, Nos. 14, 15, 16), and five landmarks from the cornus (Figure 3) were digitized.

(**a**)   (**b**)

Figure 1. Description of the landmarks used in geometric morphometric analysis. (**a**) Locations of the 20 landmarks on the forewing of *Sirex* considered in the geometric morphometric analysis, locations of the 11 landmarks on the hindwing of *Sirex* considered in the geometric morphometric analysis; (**b**) The black wing image which was converted to binary.

Figure 2. Position of 9 type landmarks on the ovipositor of *Sirex* considered in the geometric morphometric analysis.

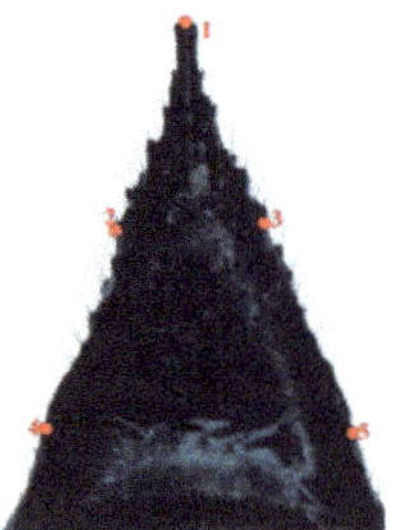

Figure 3. Position of 5 type landmarks on the cornus of *Sirex* considered in the geometric morphometric analysis.

After the. Tps files were converted into nts files using TpsUtil software and the marker information was saved, the images were processed with MorphoJ software [25,26]. The variations in shape were assessed by principal component analysis. To better visualize the variations in shape, we determined the average configuration of landmarks for each species. Deformation grids were used to show the variations. The relative similarity and discrimination of the species was analyzed using canonical variate analysis, which identified changes in shape using mean values of the two groups by assuming that covariate matrices were identical [27]. Canonical variate analysis is a reliable method for identifying differences among taxa. Procrustes ANOVA (Analysis of Variance) [25,28] was utilized to determine significant differences among species [29]. Furthermore, PAST software [30,31] was used to generate phenograms by cluster analysis that utilized Euclidean distances calculated from the matrix of the Procrustes shape coordinates. ImageJ software [32] was used to calculate the wing area. All images were converted into binary files, and the background was removed, which resulted in a black wing surrounded by a white space (Figure 1c). The wing outline was assessed, and minor damage to the wing outline was eliminated. The pixels per mm were calculated using a ruler of known scale, and the wing area was obtained.

3. Results

3.1. Shape Variables of the Wings in the Genera of S. noctilio and S. nitobei

3.1.1. Analysis of Female Forewings

Principal component analysis showed shape variations in *S. noctilio* and *S. nitobei* wings (Figure 4a). The results of Procrustes ANOVA explained 42.79% of the intergroup variations in *S. noctilio* and *S. nitobei* female forewings. Significant differences in the forewings were observed between the two species by principal component analysis (Figure 4a) and cluster analysis (Figure 4d). Mahalanobis distances between *S. noctilio* and *S. nitobei* female wings are significantly different in comparisons ($p < 0.05$), and Procrustes distances ($p < 0.05$) are similar (Table 2).

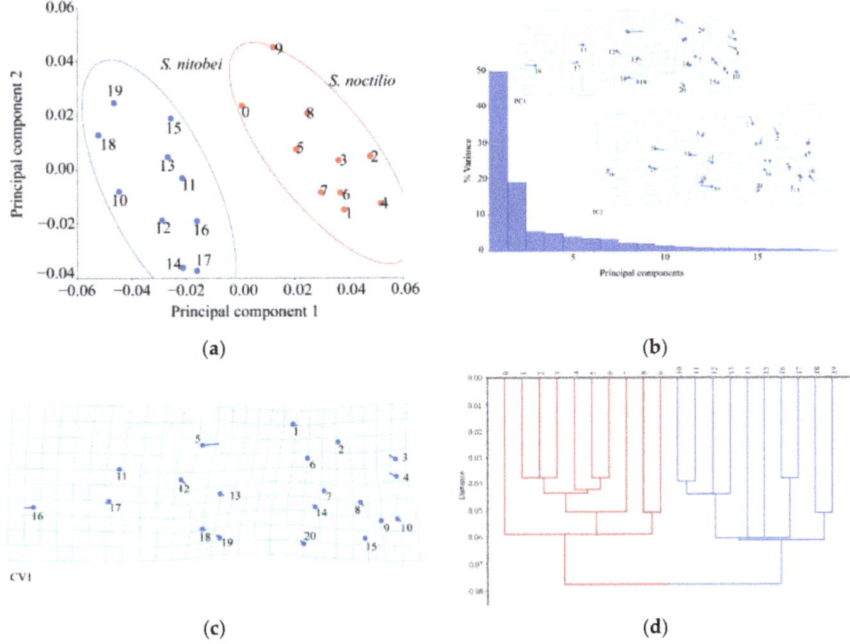

Figure 4. Shape variables of the female forewings of *S. noctilio* and *S. nitobei* (**a**) principal component analysis-(**b**) Transformation grids for visualizing a shape change (for the first two principal component, in this case)-(**c**) The Tps grids of Canonical Variate-(**d**) Phenogram of cluster analysis.

Table 2. Differences in the female wings shapes among the species of Mahalanobis distances (left) & Procrustes distances (right): *p*-values (above); distances between populations (below) (10,000 permutation rounds).

		Forewings				Hindwings			
		S. noctilio	*S. nitobei*	*S. noctilio*	*S. nitobei*	*S. noctilio*	*S. nitobei*	*S. noctilio*	*S. nitobei*
		Mahalanobis distances		Procrustes distances		Mahalanobis distances		Procrustes distances	
S. noctilio	*p*-values	-	<0.0001	-	<0.0001	-	0.0001	-	<0.0001
S. nitobei	distances	11.0555	-	0.0611	-	23.0325	-	0.0830	-

The lollipops and deformation grids indicated the direction and magnitude of the shape variations by principal component analysis (Figure 4b) and canonical variate analysis (Figure 4c). The deformation grids of the first between-group principal component revealed differences in the junction (No. 5) of Cu and 2m-cu, the junction (No. 16) of R1 and Rs2, the vannal region (Nos. 1, 3, 4) of 2A and 2cu-a, 2A and a, 1A and a. The deformation grids of the second between-group principal component revealed differences in the Rs and 2r (No. 19) and the region (around junctions Nos. 10, 17, 2). The deformation grids of the first between-group canonical variate showed that contributing most to the shape differences between them was the junctions of Cu and 2m-cu etc. (Nos. 5, 16, 3, 4, 12) (Figure 4c) (Table 3).

Table 3. Landmarks of forewing (according to veins nomenclature system by Ross (1937).

No.	Junctions of Veins	No.	Junctions of Veins
1	2A (anal veins) and 2cu-a (cubitoanal crossvein)	11	M and 3r-m (radiomedial crossvein)
2	1A and 2A	12	M and 2m-cu
3	2A and a (anal crossvein)	13	M and 2r-m
4	1A and a	14	M and 1m-cu
5	Cu (cubitus) and 2m-cu (mediocubital crossveins)	15	Rs (radial sector) and M
6	Cu and 2cu-a	16	R1 (radius) and Rs2
7	Cu and 1m-cu	17	Rs2 and 3r-m
8	Cu and 1cu-a	18	Rs and 2r-m
9	M (media) and Cu	19	Rs and 2r (radial crossvein)
10	M and Cu1	20	Rs and 1r

3.1.2. Analysis of Female Hindwings

The results of Procrustes ANOVA explained 43.71% of the intergroup variations in *S. noctilio* and *S. nitobei* female hindwings. There was also significant difference between two groups in principal component analysis (Figure 5a) and cluster analysis (Figure 5d). The deformation grids of the first principal component revealed differences in the junction (No. 2) of 1A and 2A (Figure 5b). Also, the junctions of Cu and m-cu, Cu and cu-a, C (costa) and R1 (Nos. 3, 4, 11) appeared variable in species, whereas those of the second principal component revealed differences in the remigium (Nos. 7, 5, 11, 2, 8, 10) (Figure 5b). The deformation grids of the first canonical variate showed significant differences in the region of vannal fold (Nos. 2, 3, 4) (Figure 5c) (Table 4).

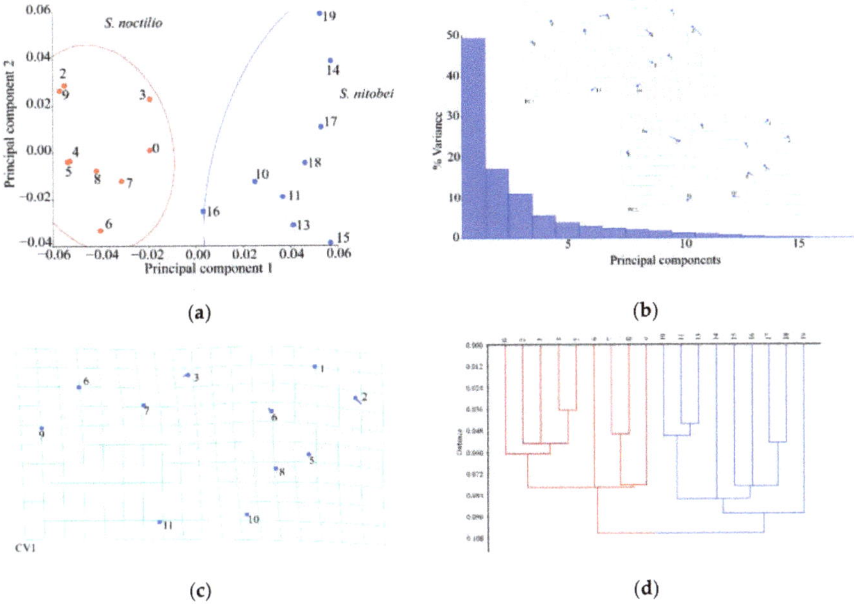

Figure 5. Shape variables of the female hindwings of *S. noctilio* and *S. nitobei* (**a**) principal component analysis-(**b**) Transformation grids for visualizing a shape change (for the first two principal component, in this case)-(**c**) The Tps grids of Canonical Variate-(**d**) Phenogram of cluster analysis.

Table 4. Landmarks of hindwing (according to veins nomenclature system by Ross (1937).

No.	Junctions of Veins	No.	Junctions of Veins
1	1A and cu-a	7	M and m-cu
2	1A and 2A	8	M and 1r-m
3	Cu and m-cu	9	Rs and 2r-m
4	Cu and cu-a	10	Sc (subcosta) and Rs
5	M and Cu	11	C (costa) and R1
6	M and 2r-m		

3.1.3. Analysis of Male Forewings

The results of Procrustes ANOVA explained 35.29% of the intergroup variations in *S. noctilio* and *S. nitobei* male forewings. Significant differences in Mahalanobis distances of the male wings were observed between the two species ($p < 0.0001$, Table 5). These findings were consistent with those of principal component analysis (Figure 6a) and cluster analysis (Figure 6d or Figure 7d).

Table 5. Differences in the male wings shapes among the species of Mahalanobis distances (left) & Procrustes distances (right): *p*-values (above); distances between populations (below) (10,000 permutation rounds).

		Forewings				Hindwings			
		S. noctilio	*S. nitobei*	*S. noctilio*	*S. nitobei*	*S. noctilio*	*S. nitobei*	*S. noctilio*	*S. nitobei*
		Mahalanobis distances		Procrustes distances		Mahalanobis distances		Procrustes distances	
S. noctilio	*p*-values	-	<0.0001	-	<0.0001	-	<0.0001	-	<0.0001
S. nitobei	distances	10.7298	-	0.0581	-	23.4236	-	0.0828	-

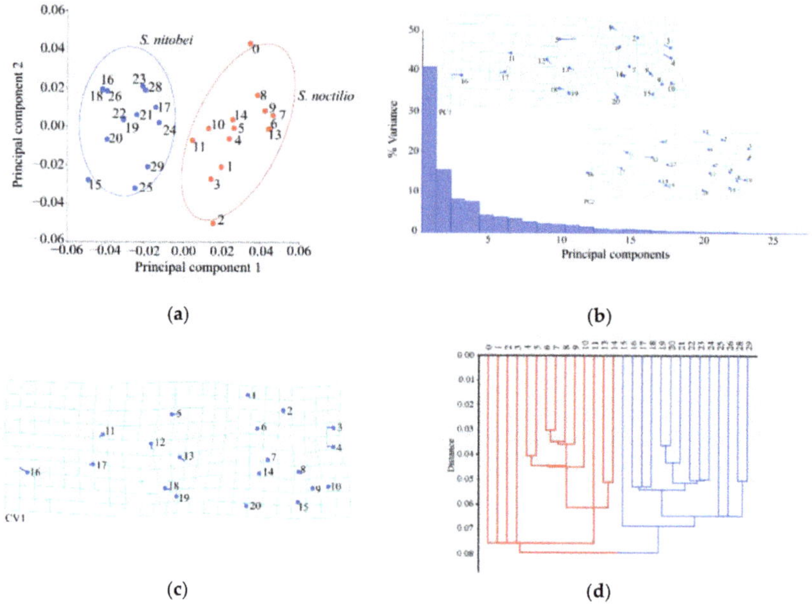

Figure 6. Shape variables of the male hindwings of *S. noctilio* and *S. nitobei* (**a**) principal component analysis-(**b**) Transformation grids for visualizing a shape change (for the first two principal component, in this case)-(**c**) The Tps grids of Canonical Variate-(**d**) Phenogram of cluster analysis.

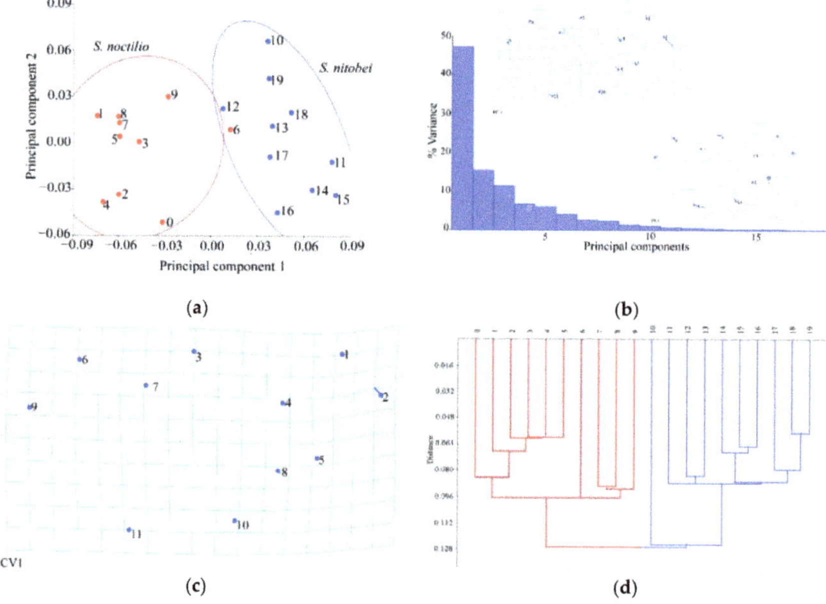

Figure 7. Shape variables of the male hindwings of *S. noctilio* and *S. nitobei* (**a**) principal component analysis-(**b**) Transformation grids for visualizing a shape change (for the first two principal component, in this case)-(**c**) The Tps grids of Canonical Variate-(**d**) Phenogram of cluster analysis.

The deformation grids of the first principal component revealed differences in the remigium (No. 16), the junction of Cu and 2m-cu, 2A and a etc. (Nos. 5, 3, 4, 18,13) (Figure 6b). These findings were similar to the deformation grids of the first canonical variate (Figure 6c). The deformation grids of the second principal component revealed differences in the junction of M and 2m-cu, 2A and 2cu-a, M and Cu1 (Nos. 12, 1, 10) (Figure 6b).

3.1.4. Analysis of Male Hindwings

The results of Procrustes ANOVA explained 32.89% of the intergroup variations in *S. noctilio* and *S. nitobei* male hindwings. The principal component analysis of the two species had individual sample overlap.

The deformation grids of the first principal component revealed differences in the region (around by junctions Nos. 2, 4, 9, 8) (Figure 7b) were similar to the deformation grids of the first canonical variate (Figure 7c), whereas those of the second principal component revealed differences in the junctions of 1A and 2A, M and 1r-m, Cu and cu-a (Figure 7b).

3.1.5. The Relationship between Wings and Dry Weight of Sirex

No significant differences were observed in the hindwings of the two woodborers. The total forewing area of male and female *S. noctilio* adults was significantly different from that of *S. nitobei* adults (F = 19.12; df = 3, 36; $p < 0.0001$; Figure 8). There was a positive correlation between the dry weight and forewing length between the two species (*S. noctilio*: r = 0.8588; $p < 0.0001$; *S. nitobei*: r = 0.8837; $p < 0.0001$; Figure 9).

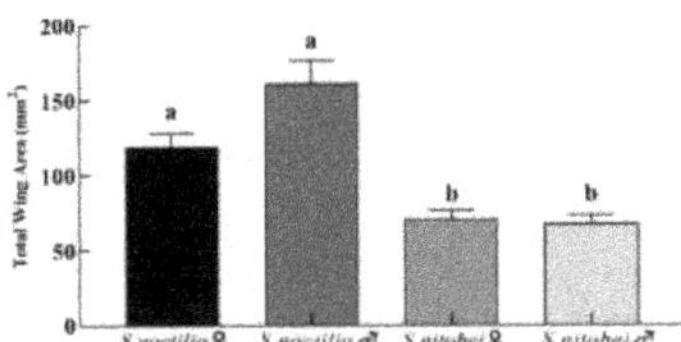

Figure 8. Comparison of total forewing area among two *Sirex*. Different letters indicate significant differences in total wing area among woodwasps within each sex, based on Tukey–Kramer's multiple comparison tests at the 5% significance level.

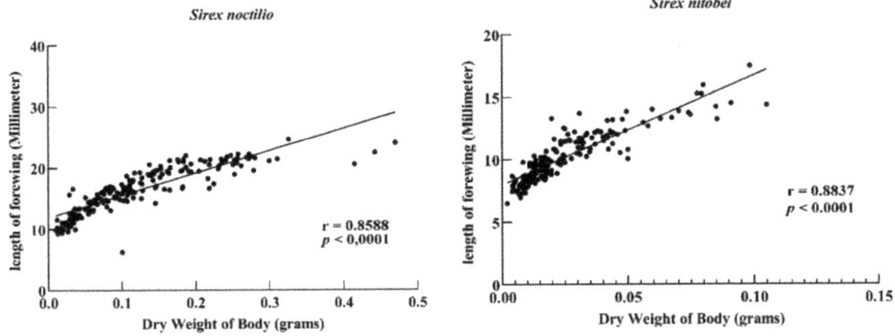

Figure 9. Relationship between the dry weight of body and the length of forewing (left: *Sirex noctilio*; right: *Sirex nitobei*).

3.2. *Shape Variables of the Ovipositors in the Genera of S. noctilio and S. nitobei*

The results of Procrustes ANOVA explained 67.94% of the intergroup variations in *S. noctilio* and *S. nitobei* female ovipositors. Mahalanobis distances between the two species were significantly different in pairwise comparisons ($p < 0.0001$), and Procrustes distances were similar ($p < 0.0001$)

(Table 6). These findings were confirmed by the results of principal component analysis (Figure 10a) and cluster analysis.

Table 6. Differences in the ovipositor shapes among the species of Mahalanobis distances (left) & Procrustes distances (right): *p*-values (above); distances between populations (below) (10,000 permutation rounds).

		Ovipositor			
		S. noctilio	*S. nitobei*	*S. noctilio*	*S. nitobei*
		Mahalanobis distances		Procrustes distances	
S. noctilio	*p*-values	-	<0.0001	-	<0.0001
S. nitobei	distances	12.0359	-	0.0986	-

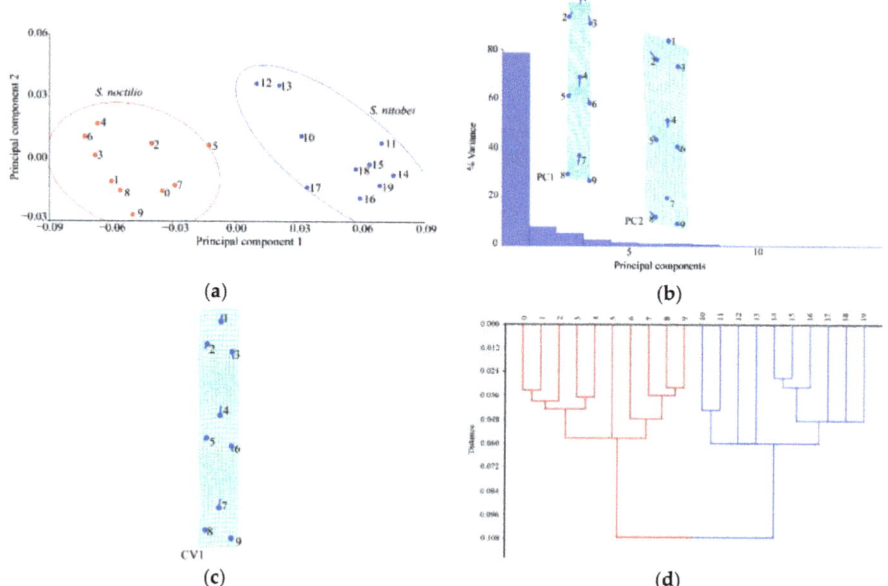

(a)   (b)

(c)   (d)

Figure 10. Shape variables of the ovipositor of *S. noctilio* and *S. nitobei* (**a**) principal component analysis-(**b**) Transformation grids for visualizing a shape change (for the first two principal components, in this case)-(**c**) The Tps grids of Canonical Variate-(**d**) Phenogram of cluster analysis.

The deformation grids of the first principal component (Figure 10b) and first canonical variate (Figure 10c) revealed differences in the angle (1, 4, 7) of the average pit, whereas those of the second between-group principal component revealed differences in the bottom left points (2, 8).

3.3. Shape Variables of the Cornus in the Genera of S. noctilio and S. nitobei

The results of Procrustes ANOVA explained 31.24% of the intergroup variations in *S. noctilio* and *S. nitobei* cornus. Mahalanobis distances among the two species were significantly different in pairwise comparisons ($p < 0.05$), and Procrustes distances were similar ($p < 0.05$) (Table 7). These findings had differences in those of cluster analysis (Figure 11d). Cluster analysis (Figure 11d) revealed that several individuals (e.g., 9, 10, 11) could not be clustered, which might have been due to differences in the tunnel environment and ossification structure. In general, the results of quantitative geometric analysis were consistent with those obtained by the naked eye.

Table 7. Differences in the cornus shapes among the species of Mahalanobis distances (left) & Procrustes distances (right): p-values (above); distances between populations (below) (10,000 permutation rounds).

		Cornus			
		S. noctilio	*S. nitobei*	*S. noctilio*	*S. nitobei*
		Mahalanobis distances		Procrustes distances	
S. noctilio	*p*-values	-	0.0067	-	0.0006
S. nitobei	distances	1.6984	-	0.1390	-

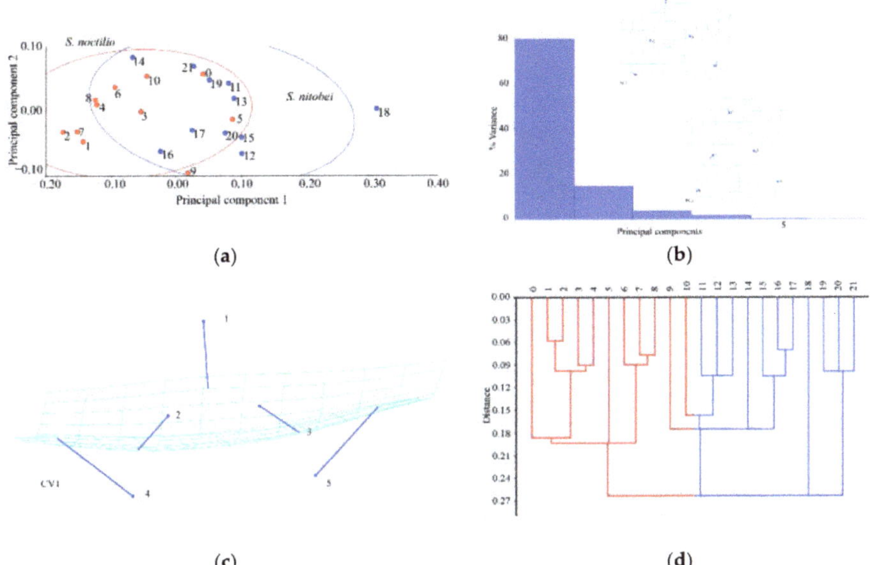

Figure 11. Shape variables of the cornus of *S. noctilio* and *S. nitobei* (**a**) principal component analysis-(**b**) Transformation grids for visualizing a shape change (for the first two principal components, in this case)-(**c**) The Tps grids of Canonical Variate-(**d**) Phenogram of cluster analysis.

4. Discussion

In recent years, we have witnessed monumental improvements in geometric morphometrics, which have enhanced the study of insect morphology. In general, these methods separate species shape from size and primarily focus on the shape as the key morphological characteristic, as few variables can reveal morphological differences between similar species. Geometric measurement methods, together with relevant mathematical models (e.g., principal component analysis, canonical variate analysis, and cluster analysis), can be used to acquire information on the morphology, genetic differentiation, development, and behavior of insects. When competing for resources on the same part of the host plant, the morphological structure of the two woodwasps may occur adaptive genetic variation. The use of geometric morphometrics allowed us to explain morphological similarities and differences between the two *Sirex* species, of which thin-spline analysis plots showed average contour distortion, and the length of the stick demonstrated the change size.

Compared with conventional classification methods, geometric survey methods can identify subtle morphological differences in insects [33]. To distinguish the two species from the traditional taxonomic morphology, we mainly distinguish the color of the male's abdomen and the female's leg. However, there is intraspecific variation in color patterns on the legs, abdomen and antennae, for example, females of *Sirex californicus*, *S. nitidus*, and *S. noctilio* each have pale and dark leg color morphs [34,35]. The results of relative warp analysis can show differences in the classification of intraspecies and interspecies [36]. In studies of morphology, the small unit on the wing is usually independent unit in the shape change, and has a certain genetic basis. Thus, different insects have different wing types and wing vein structures [3]. However, in similar species, these differences may be indistinguishable, with minor variations in the direction and branching of veins [37]. The shapes and vein profiles of insect wings contain valuable information, although it is difficult to understand the effects of behavioral and environmental factors on morphological variations using conventional classification methods [33].

Insect flight involves the first two branches of the radius vein and the thicker area of the wing film [38]), which is gradually reduced from the base to the end of the wing. *S. noctilio* woodwasps have a variable flight behavior, which relates to initial body size [39]. The body size of the *Sirex* also varies in different regions, and the two woodwasps cannot be distinguished simply by it. *Sirex* species mainly use carbohydrates for fuel during flight. The dry weight of *S. noctilio* is significantly larger than *S. nitobei* in this study. Therefore, the wing variations of both *Sirex* species in this study might explain the differences in flight ability and behavior. In addition, the ovipositor, an appendage through which females deposit eggs, plays important roles in sensing the microenvironment and initiating the laying of eggs. In this study, we observed differences in forewing and ovipositor shape, which were due to developmental plasticity. These findings provide a strong basis for further research on flight behavior and ovipositor function in various species.

In this study, we identified and classified two *Sirex* species, although the lack of inclusion of other Siricidae species was a major limitation. When species cannot be immediately identified by their appearance, landmark points can be easily extracted and analyzed. We anticipate that additional landmark points, such as those of the head, chest, and other parts, will be used in future studies [40]. Geographic distance is one of the key factors of population differentiation for widespread species. It is generally believed that the more geographically separated populations have less chance of gene exchange, resulting in morphological differences between populations. *S. noctilio* has invaded many areas in China, next we can collect samples from different regions for analysis of different geographical populations. At the same time, this paper provides basic data for the automatic recognition system of insects. Further identification of *Sirex* species will be provided. In addition, insects have a short life cycle and a fast response to the environment. Using geometric morphology can accurately analyze small changes in morphological structures and possible evolutionary trends in a short period of time.

5. Conclusions

There is species variation in the wing veins of two woodwasps. We selected the homology coordinate points for analysis. In conclusion, there were significant differences in the forewings and the pits on the ovipositor between the invasive *S. noctilio* and the native *S. nitobei*. The results showed that the taxonomic importance of hind wing venation and cornus characters was not stable for the two woodwasps. Geometric morphology can be used for morphological identification of insects of the genus *Sirex*, especially those species with variable coloration. We can distinguish the two woodwasps from the flight-related forewing veins and the reproduction-related ovipositor pits. Comparing the invasive species with their congeners can partialy avoid the bias due to taxonomical relatedness and enhance the credibility of the results. With landmark-based geometric morphometrics to quantify and analyze wing, cornus, and ovipositor morphologies of two *Sirex* species, we provide new insights for automatic insect identification systems. The currently used approach to study the morphology of wings is complicated and time consuming. This process may damage wings and several software packages cannot extract landmark points. In these cases, the user must use a computer mouse to manually select landmarks, as was performed in this study; however, measurements are affected by human factors. Further studies are needed to expand the identification system of insects by including different types of insects and performing different types of geometric morphometric analysis. New insect identification software packages should also be developed, as they can reduce the repetitive workload for investigators working in agriculture, forestry, quarantine, and other front-line industries.

Author Contributions: Conceptualization, L.R. and Y.L.; data curation, M.W.; formal analysis, M.W.; funding acquisition, L.R. and Y.L.; methodology, M.W., L.R., and Y.L.; project administration, L.R. and Y.L.; resources, M.W., L.W., N.F., C.G. and T.A.; supervision, L.R. and Y.L.; writing—original draft, M.W.; writing—review and editing, M.W. and L.R. All authors have read and agreed to the published version of the manuscript.

Funding: This work was supported by the National Key Research & Development Program of China "Research on key technologies for prevention and control of major disasters in plantation" (2018YFD0600200), and Beijing's Science and Technology Planning Project "Key technologies for prevention and control of major pests in Beijing ecological public welfare forests" (Z191100008519004).

Acknowledgments: We greatly appreciated the help from Jinbaotun Forest station in Tongliao City, Inner Mongolia and workers of the Forestry Bureau for their assistance with fieldwork.

Conflicts of Interest: The authors declare no potential conflict of interest.

References

1. Xiao, G.R.; Wu, J. The Siricid wood wasps of China (Hymenoptera, Symphyta). *Sci. Silvae Sin.* **1983**, *19*, 1–29.
2. Pretorius, E. Using geometric morphometrics to investigate wing dimorphism in males and females of hymenoptera—A case study based on the genus *Tachysphex kohl* (Hymenoptera: Sphecidae: Larrinae). *Aust. J. Entomol.* **2005**, *44*, 113–121. [CrossRef]
3. Zimmerman, E.; Palsson, A.; Gibson, G. Quantitative trait loci affecting components of wing shape in *Drosophila melanogaster*. *Genetics* **2000**, *155*, 671–683. [CrossRef]
4. Tabugo, S.R.M.; Torres, M.A.J.; Demayo, C.G. Determination of developmental modules and conservatism in the fore- and hind-wings of two species of dragonflies, *Orthetrum sabina* and *Neurothemis ramburii*. *Int. J. Agric. Biol.* **2011**, *13*, 541–546. [CrossRef]
5. Daly, V.H. Insect morphometrics. *Annu. Rev. Entomol.* **1985**, *30*, 415–438. [CrossRef]
6. Rohlf, F.J.; Marcus, L.F. A revolution in morphometrics. *Trends Ecol. Evol.* **1993**, *8*, 129–132. [CrossRef]
7. Marcus, L.F.; Corti, M.; Loy, A.; Naylor, G.J.P.; Slice, D.E. *Advances in Morphometrics*; Springer: New York, NY, USA, 1996. [CrossRef]
8. Adams, D.C.; Rohlf, F.J.; Slice, D.E. Geometric morphometrics: Ten years of progress following the 'revolution'. *Ital. J. Zool.* **2004**, *71*, 5–16. [CrossRef]
9. Pan, P.L.; Shen, Z.R.; Gao, L.W.; Yang, H.Z. Development of the technology for auto-extracting venation of insects. *Entomotaxonomia* **2008**, *30*, 72–80. [CrossRef]
10. Dujardin, J.P.; Le Pont, F.; Baylac, M. Geographical versus interspecific differentiation of sand flies (Diptera: Psychodidae): A landmark data analysis. *Bull. Entomol. Res.* **2003**, *93*, 87–90. [CrossRef]
11. Baylac, M.; Villemant, C.; Simbolotti, G. Combining geometric morphometrics with pattern recognition for the investigation of species complexes. *Biol. J. Linn. Soc.* **2003**, *80*, 89–98. [CrossRef]
12. De la Rúa, P.; Galian, J.; Serrano, J.; Moritz, R.F.A. Genetic structure and distinctness of *Apis mellifera* L. populations from the Canary Islands. *Mol. Ecol.* **2001**, *10*, 1733–1742. [CrossRef]
13. Klingenberg, C.P. Novelty and "homology-free" morphometrics: What's in a name? *Evol Biol.* **2008**, *35*, 186–190. [CrossRef]
14. Hurley, B.P.; Slippers, B.; Wingfield, M.J. A comparison of control results for the alien invasive Woodwasp, *Sirex noctilio*, in the Southern Hemisphere. *Agric. For. Entomol.* **2007**, *9*, 159–171. [CrossRef]
15. Slippers, B.; De Groot, P.; Wingfield, M.J. *The Sirex Woodwasp and Its Fungal Symbiont*; Springer: Dordrecht, The Netherlands, 2012. [CrossRef]
16. Li, D.P.; Shi, J.; Lu, M.; Ren, L.L.; Zhen, C.; Luo, Y.Q. Detection and identification of the invasive *Sirex noctilio* (Hymenoptera: Siricidae) fungal symbiont, *Amylostereum areolatum* (Russulales: Amylostereacea), in China and the stimulating effect of insect venom on laccase production by *A. areolatum* YQL03. *J. Econ. Entomol.* **2015**, *108*, 1136–1147. [CrossRef]
17. Wang, M.; Bao, M.; Ao, T.G.; Ren, L.L.; Luo, Y.Q. Population distribution patterns and ecological niches of two *Sirex* species damaging *Pinus sylvestris* var. *mongolica*. *Chin. J. Appl. Entomol.* **2017**, *54*, 924–932. [CrossRef]
18. Bai, M.; Yang, X.K.; Li, J.; Wang, W.C. Geometric morphometrics, a super scientific computing tool in morphology comparison. *Chin. Sci. Bull.* **2014**, *59*, 887–894. [CrossRef]

19. Bookstein, F.L. Morphometric Tools for Landmark Data. Cambridge University: Cambs, UK, 1992; Volume 87, pp. 55–87. [CrossRef]

20. Bookstein, F.L. Combining the tools of geometric morphometrics. In *Advances in Morphometrics*; NATO ASI Series (Series A: Life Sciences); Marcus, L.F., Corti, M., Loy, A., Naylor, G.J.P., Slice, D.E., Eds.; Springer: Boston, MA, USA; New York, NY, USA, 1996; Volume 284, pp. 131–152. [CrossRef]

21. Hanken, J. *Book Review: Morphometrics in Evolutionary Biology. The Geometry of Size and Shape Change, with Examples From Fishes*, 15th ed.; Bookstein, F.L., Chernoff, B., Elder, R.L., Humphries, J.M., Smith, J.G.R., Strauss, R.E., Eds.; The University of Chicago Press: Chicago, IL, USA, 1986; Volume 61, pp. 542–543. [CrossRef]

22. Bookstein, F.L. Morphometric tools for landmark data; Geometry and biology. *Q. Rev. Biol.* **1993**, *54*, 171–173. [CrossRef]

23. Sun, X.T.; Tao, J.; Ren, L.L.; Shi, J.; Luo, Y.Q. Identification of\r, Sirex noctilio\r, (Hymenoptera: Siricidae) using a species-specific cytochrome c oxidase subunit I PCR Assay. *J. Econ. Entomol.* **2016**, *109*, 1424–1430. [CrossRef]

24. Rohlf, F.J. *Tps Series*; Department of Ecology and Evolution, State University of New York: Stony Brook, New York, NY, USA, 2010; Available online: http://life.bio.sunysb.edu/morph/ (accessed on 8 June 2011).

25. Klingenberg, C.P.; Barluenga, M.; Meyer, A. Shape analysis of symmetric structures: Quantifying variation among individuals and asymmetry. *Evolution* **2002**, *56*, 1909–1920. [CrossRef]

26. Klingenberg, C.P. MorphoJ: An integrated software package for geometric Morphometrics. *Mol. Ecol. Resour.* **2011**, *11*, 353–357. [CrossRef]

27. Klingenberg, C.P. Evolution and development of shape: Integrating quantitative approaches. *Nat. Rev. Genet.* **2010**, *11*, 623–635. [CrossRef]

28. Klingenberg, C.P.; McIntyre, G.S. Geometric morphometrics of devel- opmental instability: Analyzing patterns of fluctuating asymmetry with Procrustes methods. *Evolution* **1998**, *52*, 1363–1375. [CrossRef] [PubMed]

29. Klingenberg, C.P.; Monteiro, L.R. Distances and directions in multidimensional shape spaces: Implications for morphometric applications. *Syst. Biol.* **2005**, *54*, 678–688. [CrossRef] [PubMed]

30. Hammer, Ø.; Harper, D.A.T.; Ryan, P.D. PAST: Paleontological statistics software package for education and data analysis. *Paleontol. Electron.* **2001**, *4*, 3–9.

31. Hammer, Ø. PAST Ver. 2.09. Available online: http://norges.uio.no/past/download.html (accessed on 8 June 2011).

32. Ferreira, T.; Rasband, W.S. ImageJ User Guide -IJ/1.46. Available online: imagej.nih.gov/ij/docs/guide/ (accessed on 26 June 2012).

33. Lawing, A.M.; Polly, P.D. Geometric morphometrics: Recent applications to the study of evolution and development. *J. Zool.* **2010**, *280*, 1–7. [CrossRef]

34. Schiff, N.M.; Goulet, A.H.; Smith, D.R.; Boudreault, C.; Wilson, A.D.; Scheffler, B.E. Siricidae (Hymenoptera: Symphyta: Siricoidea) of the Western Hemisphere. *Can. J. Arthropod Identif.* **2011**, *21*, 1–305.

35. Ross, H.H. A generic classifcation of the Nearctic sawflies (Hymenoptera: Symphyta). *Ill. Biol. Monogr.* **1937**, *15*, 1–173.

36. Kendall, D.G.; Le, H.L. Exact shape-densities for random triangles in convex polygons. *Adv. Appl. Probab.* **1986**, *18*, 59–72. [CrossRef]

37. Shevtsova, E.; Hansson, C.; Janzen, D.H.; Kjaerandsen, J. Stable structural color patterns displayed on transparent insect wings. *Proc. Natl. Acad. Sci. USA* **2011**, *108*, 668–673. [CrossRef]

38. Combes, S.A.; Daniel, T.L. Flexural stiffness in insect wings. II. Spatial distribution and dynamic wing bending. *J. Exp. Biol.* **2003**, *206*, 2989–2997. [CrossRef]

39. Bruzzone, O.A.; Villacide, J.M.; Bernstein, C.; Corley, J.C. Flight variability in the woodwasp *Sirex noctilio* (Hymenoptera: Siricidae): An analysis of flight data using wavelets. *J. Exp. Biol.* **2009**, *212*, 731–737. [CrossRef] [PubMed]

40. Yang, H.Z.; Cai, X.N.; Li, X.T.; Sun, Z.R. Application of geometric morphometrics in insect identification. *Sichuan J. Zool.* **2013**, *32*, 464–469. [CrossRef]

 *insects*

Article

WingMesh: A Matlab-Based Application for Finite Element Modeling of Insect Wings

Shahab Eshghi [1,*], **Vahid Nooraeefar** [2], **Abolfazl Darvizeh** [2], **Stanislav N. Gorb** [1] and **Hamed Rajabi** [1]

1   Functional Morphology and Biomechanics, Institute of Zoology, Kiel University, 24118 Kiel, Germany; sgorb@zoologie.uni-kiel.de (S.N.G.); harajabi@hotmail.com (H.R.)
2   Faculty of Mechanical Engineering, University of Guilan, Rasht 4199613776, Iran; vahid.nooraeefar@gmail.com (V.N.); adarvizeh@guilan.ac.ir (A.D.)
*   Correspondence: eshghi.shahab@gmail.com

Received: 20 July 2020; Accepted: 18 August 2020; Published: 18 August 2020

Simple Summary: Manual modeling of complicated insect wings presents considerable practical challenges. To overcome these challenges, therefore, we developed *WingMesh*. This is an application for simple yet precise automatic modeling of insect wings. Using a series of examples, we showed the performance of our application in practice. We expect *WingMesh* to be particularly useful in comparative studies, especially where the modeling of a large number of insect wings is required within a short time.

Abstract: The finite element (FE) method is one of the most widely used numerical techniques for the simulation of the mechanical behavior of engineering and biological objects. Although very efficient, the use of the FE method relies on the development of accurate models of the objects under consideration. The development of detailed FE models of often complex-shaped objects, however, can be a time-consuming and error-prone procedure in practice. Hence, many researchers aim to reach a compromise between the simplicity and accuracy of their developed models. In this study, we adapted *Distmesh2D*, a popular meshing tool, to develop a powerful application for the modeling of geometrically complex objects, such as insect wings. The use of the burning algorithm (BA) in digital image processing (DIP) enabled our method to automatically detect an arbitrary domain and its subdomains in a given image. This algorithm, in combination with the mesh generator *Distmesh2D*, was used to develop detailed FE models of both planar and out-of-plane (i.e., three-dimensionally corrugated) domains containing discontinuities and consisting of numerous subdomains. To easily implement the method, we developed an application using the Matlab App Designer. This application, called *WingMesh*, was particularly designed and applied for rapid numerical modeling of complicated insect wings but is also applicable for modeling purposes in the earth, engineering, mathematical, and physical sciences.

Keywords: biological structures; computer vision; mesh generation; simulation; digital image processing

1. Introduction

The finite element (FE) method is a numerical technique which is generally used to simulate a physical phenomenon in the virtual world by solving complex boundary value problems [1,2]. FE software packages were developed to simplify often complicated simulation processes. They are especially very common in engineering applications [3–5] and are becoming increasingly popular in the investigation of the mechanical behavior of biological structures, such as complex human and animal body parts [6–13].

Although providing a user with a high degree of flexibility, all available FE packages have a common need: an accurate model. A model is a domain which is subdivided into smaller polygonal or polyhedral meshes, so-called elements [14]. A modeling process, however, may present many challenges and can be rather time-consuming, especially when dealing with complex geometries [15–17], which is usually the case in biology. The skills of the software user can also strongly influence the process and the final result. These often lead to oversimplified models and, therefore, can affect the accuracy of simulation results. How can this problem be overcome?

In 2004, Persson and Strang aimed to address this problem by developing a simple meshing technique called *Distmesh2D* [18]. As intended by its developers, the method, which was implemented in Matlab code, provided an effective tool to mesh a given domain automatically. The simplicity of the method and the high quality of the produced mesh are the key advantages of the proposed method. However, it also has a major drawback: finding the distance to boundaries by the use of the mathematical equation $f(x, y) = 0$ or by values of a discrete set of points, as explained by the authors, is a time-consuming and error-prone procedure for complex geometries. Due to the use of a mathematical scheme to define the distance function (see Section 2.5), *Distmesh2D* also has limitations when meshing a domain containing discontinuities.

Here, we aimed to address these challenges and improve the performance of *Distmesh2D* but still maintain its simplicity. To this end, we used computer vision to automatically detect the boundary of a domain in a given image. We combined it with the mesh generator *Distmesh2D* to develop an application for the rapid modeling of geometrically complex domains that consist of several subdomains. The applicability of the method is not limited to in-plane domains, but it can also mesh out-of-plane (i.e., corrugated) objects. We specifically designed and used our method to develop models of insect wings. The proposed application, called *WingMesh*, draws extensively on Persson and Strang's account in an attempt to offer a simple, but more practical, meshing tool.

2. Materials and Methods

The modeling method presented in this study, called *WingMesh*, consists of several algorithms that interact with each other. The method requires an input image to identify the boundaries of a given domain as well as subdomains within that. The code *Distmesh2D* is then employed to mesh the identified domain. Other algorithms were added to the main algorithms to model out-of-plane domains and create a **.inp* file, which is the Abaqus input file format [19]. The key parts of the developed Matlab code and the full code of the method, together with a graphical user interface (GUI) (see Section 3), are accessible in the Supplementary Materials (Codes S1 and S3, Method S1).

2.1. Burning Algorithm for Detection of the Boundary of a Given Domain

The burning algorithm (BA) was used to extract the boundary of a domain in a given image [20,21]. The BA needs a digital black and white image as the input, in which black pixels, with the pixel value of 0, represent the border of the domain and white pixels, with the pixel value of 1, represent regions that are situated inside and outside of the domain. The BA uses the matrix of the input image to detect the boundary of the domain in that image. This process starts with choosing a pixel within the domain by the user (Pixel 1 in Figure 1a) and continues by detecting white pixels around the selected pixel. To this end, the algorithm checks the colors of pixels located in the four orthogonal directions of the selected white pixels (Pixels 2 in Figure 1b). The coordinates of the found black pixels are stored in a matrix, and the colors of detected white and black pixels are changed to 0.8 (light grey) and 0.1 (dark grey), respectively, in order to avoid their reselection in the next iteration. This process continues by searching for white and black pixels around only white pixels detected in the previous iteration (Figure 1c–m). This process continues until all white pixels inside the domain are detected (Figure 1m). Code S2 and Video S1 in the Supplementary Materials are the source code of BA and a simple illustration of how it works, respectively.

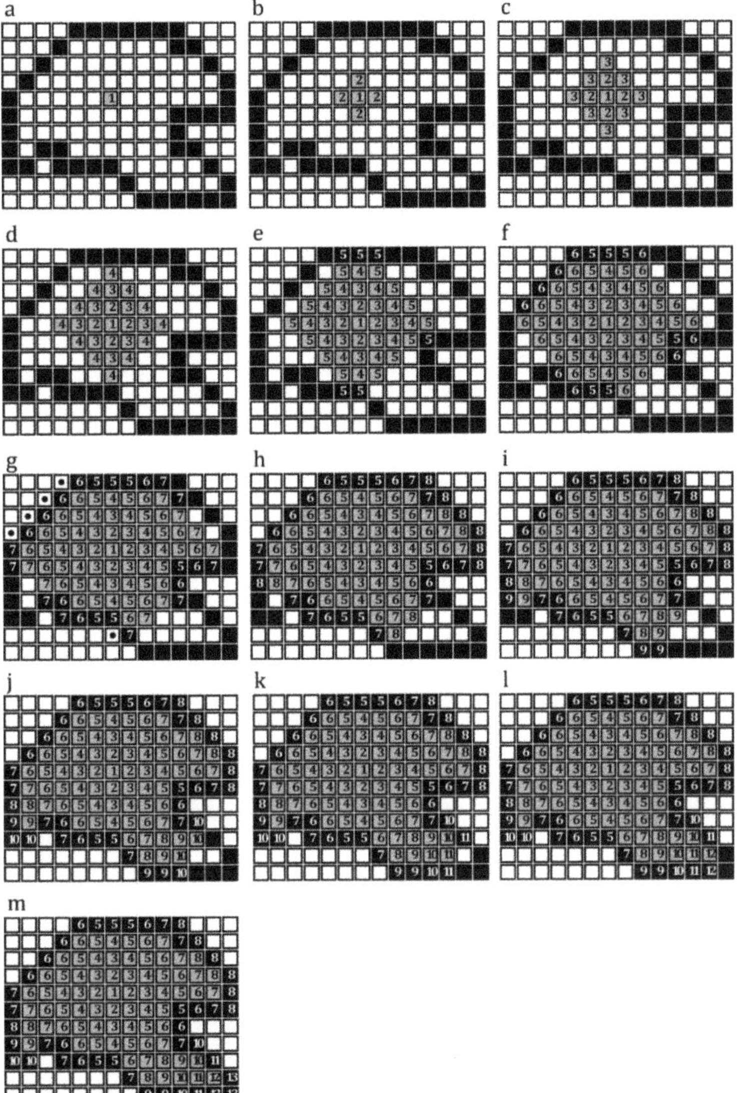

Figure 1. Detection of the border of an arbitrary domain using the Burning Algorithm. (**a**) A white pixel inside the domain is selected. (**b**) The BA searches for white and black pixels around the selected pixel in four orthogonal directions. (**c**) The BA searches for white and black pixels around the detected white pixels in the previous iteration. (**d**–**m**) This process continues until there is no white pixel inside the domain (**m**).

2.2. Detection of Subdomains within a Given Domain

The function BA can also detect subdomains within a given domain. When the main domain contains any subdomain, the application first finds the white pixels outside the domain. This process eventually results in the detection of the boundary of the domain.

To find the boundary of each subdomain, the user should select a pixel inside that subdomain in the input image. By this, the BA finds the pixels located on the boundary of the subdomain using the same method explained earlier. This process continues as long as the user selects a pixel in a new subdomain.

2.3. Detection of Discontinuities in a Given Domain

The function BA can detect any discontinuity, such as holes, cracks, etc., in a given domain. For this purpose, if any discontinuity exists, the user should select a pixel in each discontinuity in the input image (Video S2). After this, using the same method as described before, BA finds the boundary of each selected discontinuity. By this, the application detects discontinuities and excludes them from the main domain. To this end, after meshing the structure, the application finds all elements inside the region of discontinuity and excludes them from the model.

2.4. Development of a Corrugated Model

In a recent study, we developed a method for modeling out-of-plane (i.e., corrugated) domains [22]. Here, we modified this technique to make it more efficient and easier to implement. This technique requires an additional input image with the same frame size as the main input image. The other image should include information regarding the corrugation of the out-of-plane domain. The information should include the location of the maximum and minimum heights, indicated by the black and white colors, respectively. The value of pixels in the secondary image, therefore, serves as a measure of the height of that pixel: pixel values 0 and 1 indicate the maximum and minim heights, respectively. If there is more than one maximum or minimum height in a domain, any local extremum can be marked in grey color. The intensity of the grey color in each local extremum indicates the relative height of that extremum compared to the absolute extremum.

The recursive Equation (1) is used to smooth the corrugations to avoid any abrupt change in the height of a model at the location of an extremum.

$$v(r.c) = mean\left(\sum_{i=-1}^{1} \sum_{j=-1}^{1} v(r + i.c + j) \right) \tag{1}$$

where r and c represent the number of the row and column of a pixel in the image, respectively. v is the value of that pixel. i and j are the index of the row and column of the pixels in the image. This equation recursively updates the color intensity of the pixels in the secondary image and, thereby, the height of those pixels in the model. The number of iterations, which is set by the user, controls the sharpness of corrugations in the developed model. The values of pixels in the secondary input image, which are between 0 and 1, represent the relative heights of corrugations.

Figure 2a shows a corrugated object. The image of the object from the top view is shown in Figure 2b. Figure 2c shows the secondary input image, which has the same frame size as the image shown in Figure 2b. The black line in the middle of the image represents the position of the only available height maximum, and the white color corresponds to the regions with the minimum height. When using these two images, the application develops a model similar to that shown in Figure 2d. Figure 2j shows the gradual changes in the corrugation of the model by the use of Equation (1) after 20, 100, 150, 200, and 300 iterations (Figure 2e–i).

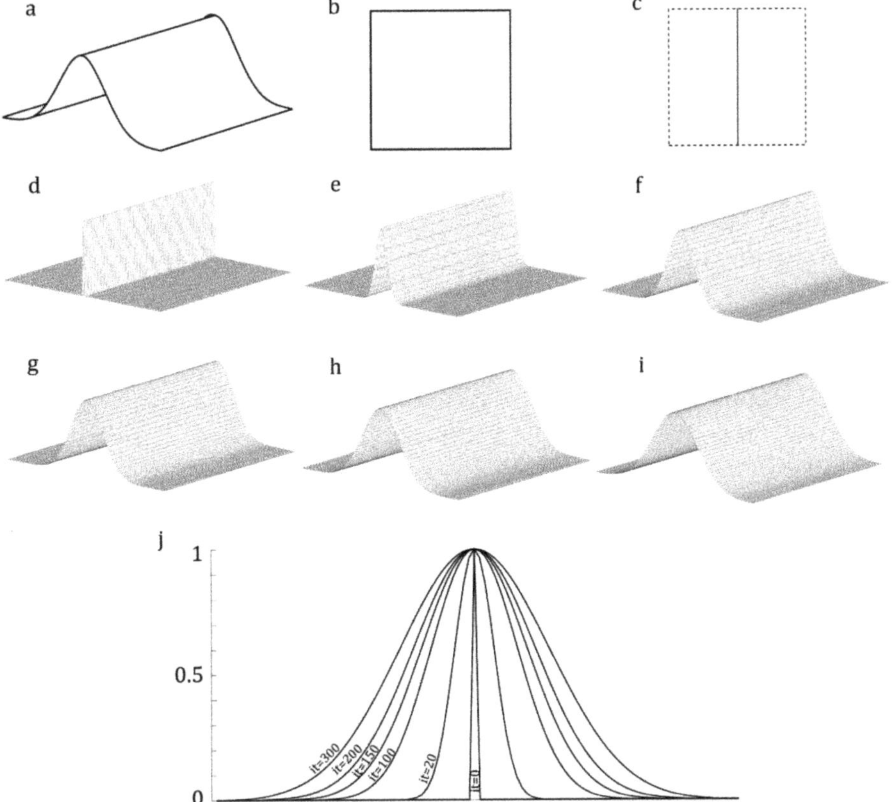

Figure 2. Modeling of an out-of-plane domain. (**a**) An out-of-plane domain. (**b**) A top view image of the domain. The image is used by the BA to detect the domain. (**c**) The secondary image contains a black line that represents the maximum height. The regions with zero height are colored in white. (**d**) A developed model based on the input images. (**e–i**) Smoothing the height of the meshed model using the iterative algorithm. (**j**) Changes in the corrugation pattern in different iterations.

2.5. Mesh Generation

Distmesh2D is a mesh generator in Matlab which employs a distance function, $d(x,y)$, to describe the geometry of a domain [18]. The Delaunay algorithm is used in *Distmesh2D* to generate triangular meshes. The first line of the code *Distmesh2D*, the calling syntax, represents inputs and outputs of the Matlab code:

$$function\ [p,t] = distmesh2d(fd,fh,h0,bbox,pfix)$$

where the input arguments are as follows:

1. *fd*, the distance function that defines the boundary of the domain.
2. *fh*, the distance function, which controls the convergence of the size of elements. The size of the elements decreases near *fh*.
3. *h0*, the distance between nodes in the initial distribution.
4. *bbox*, the bounding box in which the domain is located.
5. *pfix*, defines nodal points, which are set as fixed points while generating elements.

Distmesh2D produces the following outputs:

1. p, gives the coordinate of the nodal points.
2. t, indicates the connection between the nodes.

Here, coordinates of the nodes on the boundary of a given domain, which are obtained by the BA, are used to define the distance functions *fd* and *fh* for the mesh generator *Distmesh2D*. In addition to the distance functions *fd* and *fh*, *Distmesh2D* has three other inputs: *h0*, *bbox*, and *pfix*. *h0*, the distance between initial nodes, can be set to 1, because the minimum distance between two pixels is 1. *bbox* is equal to the frame size of the imported image (size of the input matrix). The pixels located on the boundaries of the subdomains, extracted by the BA, are defined as fixed points, *pfix*.

Distmesh2D can generate both structured and unstructured elements. However, in this study, we set it to create only unstructured elements, because this type of element fits better with our aim for developing models of geometrically complex structures.

2.6. Outputs

WingMesh generates a **.inp* file (i.e., an Abaqus input file) which contains information regarding the coordinates of the nodal points, their connections, type of elements, sections of the domain, and the material properties of sections. Detailed information about **.inp* files can be found in the Supplementary Materials (Method S1).

3. Graphical User Interface

WingMesh was coded in Matlab 2019a, and Matlab App Designer was employed to develop a GUI. This GUI makes the method easy to implement and eliminates the need to know a programming language. The GUI is available in Code S3, and its description is available in Method S1.

4. Examples

*   Example 1: An in-plane domain

Figure 3a shows a single in-plane domain with straight-line borders and sharp corners. Figure 3b shows the output model. The **.inp* output file developed by the method is available in File S1.

*   Example 2: An in-plane domain consisting of two subdomains

Figure 3c illustrates the same domain shown in Figure 3a, which is subdivided into two subdomains. As shown in Figure 3d, *WingMesh* was able to detect the border between the subdomains. The subdomains have meshed separately as two sections of a single model. The generated **.inp* file is available in File S2.

*   Example 3: An in-plane domain with subdomains and a discontinuity

We added a circular hole within one of the two subdomains of the domain given in the previous example (Figure 3e). After meshing all subdomains, including the discontinuity in the main domain, the elements generated in the discontinuity were removed before the final model was developed (Figure 3f). The **.inp* file is available in File S3.

*   Example 4: An irregular-shaped in-plane domain with several discontinuities

Figure 3g illustrates an irregular-shaped domain with curved borders, which contains four discontinuities. Previously, it was impossible to model such an irregular domain with complex-shaped discontinuities using the mesh generator *Distmesh2D*. However, the use of the DIP technique enables *WingMesh* to mesh such geometries. Figure 3h shows the meshed model developed based on the given domain. The **.inp* file is available in File S4.

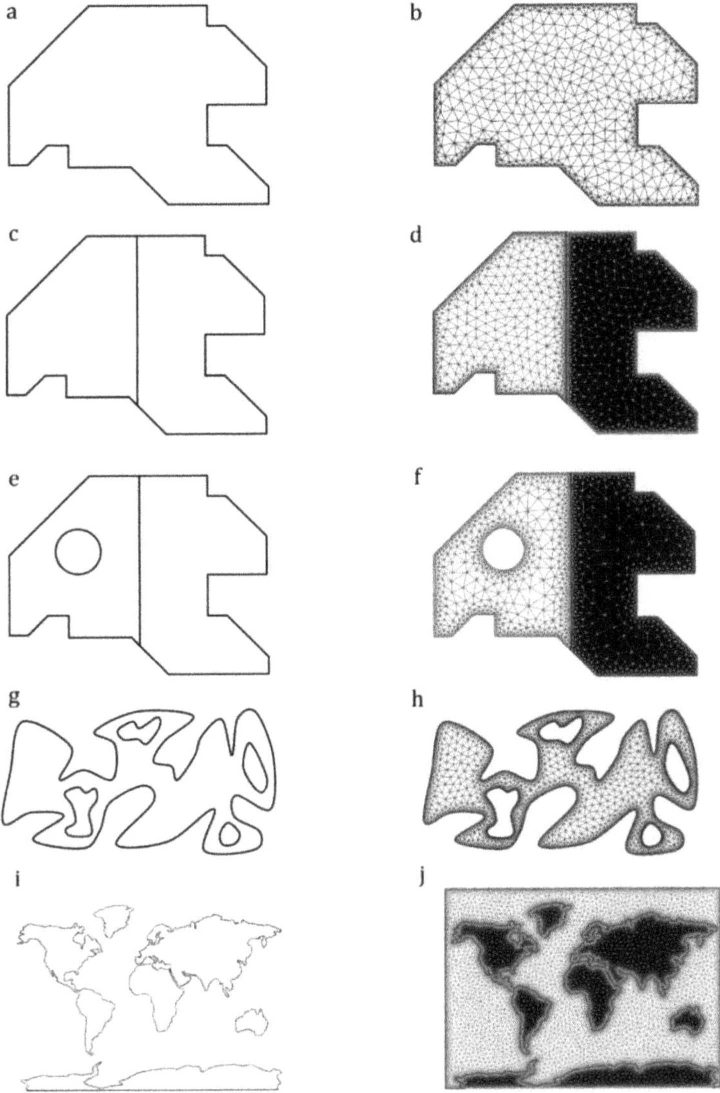

Figure 3. Modeling of in-plane domains. (**a**) Image of a simple in-plane domain. (**b**) The meshed model developed from the image in (**a**). (**c**) Image of an in-plane domain consisting of two subdomains. (**d**) The meshed model developed from the image in (**c**). (**e**) Image of an in-plane domain with two subdomains and a discontinuity. (**f**) The meshed model developed from the image in (**e**). (**g**) Image of an irregular-shaped domain with several discontinuities. (**h**) The meshed model developed from the image in (**g**). (**i**) Image of a complex-shaped in-plane domain with several subdomains. (**j**) The meshed model developed from the image in (**i**).

- Example 5: A complex-shaped in-plane domain with several subdomains

Figure 3i shows the world map with the irregular shaped continents. The meshed model, which is apparently in good agreement with the given image, is presented in Figure 3j. The **.inp* file is available in File S5.

- Example 6: An asymmetric out-of-plane domain with one height maximum and one height minimum

In this example and the next three cases, we used the domain shown in Figure 3a to develop out-of-plane models with different corrugation patterns. Here, we used the image in Figure 4a as the secondary input image to provide information on the corrugation spots. The grey color in this image indicates regions with zero height. The black and white lines indicate a height maximum and a height minimum, respectively. Figure 4b,c shows the perspective and side views of the meshed model. The *.inp* file is available in File S6.

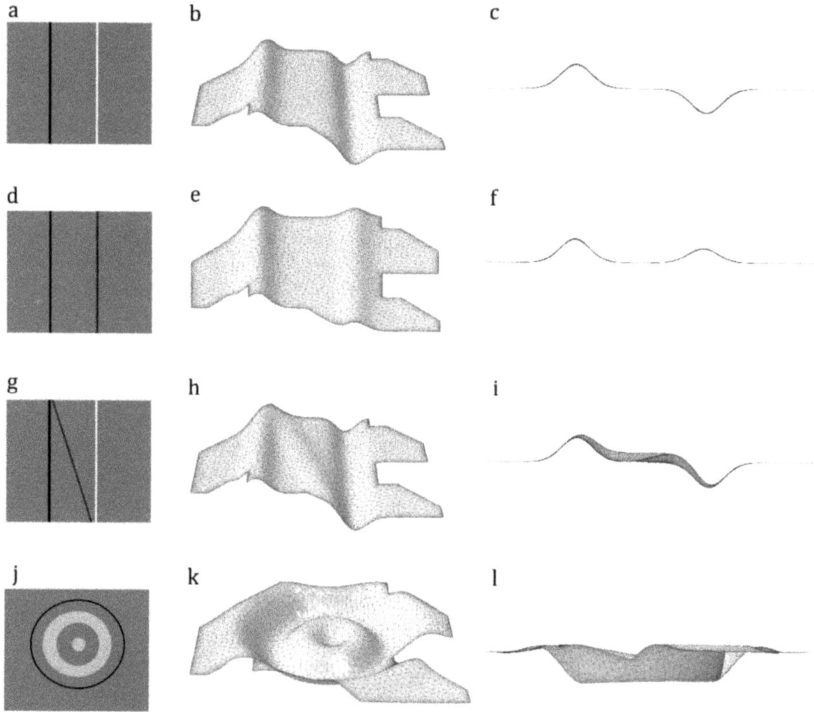

Figure 4. Modeling of out-of-plane domains. The use of different secondary images in combination with the same input image, as shown in Figure 3a, results in the development of models with different corrugated patterns. (**a,d,g,j**) Secondary images contain information on corrugation spots. (**b,e,h,k**) Perspective views of the meshed models created based on the image shown in Figure 3a and secondary images shown in Figure 4a,d,g,j. (**c,f,i,l**) Side views of meshed models.

- Example 7: An out-of-plane domain with two height maxima

Figure 4d shows an image with the black and dark grey lines, which represent two height extrema. Using this as a secondary image results in the development of the meshed model shown in Figure 4e,f. The *.inp* file is available in File S7.

- Example 8: An out-of-plane domain with two height maxima and a height minimum

In Figure 4g, we added a tilted grey line to those in the secondary image shown in Figure 4a. The grey line is expected to change the corrugation pattern of the meshed model in Figure 4b by adding a region with a height maximum. Figure 4h,i shows the perspective and side views of the model developed using the secondary image in Figure 4g. The *.inp* file is available in File S8.

- Example 9: An out-of-plane domain with circumferentially oriented height extrema

In this example, we aimed to test the precision of our method by developing a more complex corrugated domain. In this domain, the corrugation spot is circumferentially oriented, compared with the other domains that had longitudinal corrugations. Figure 4k,l shows the model developed by using Figure 4j. The *.inp* file is available in File S9.

- Example 10: A beetle wing

Figure 5a shows the hind wing of a beetle, *Allomyrina dichotoma* (Coleoptera: Scarabaeidae). Figure 5b shows the secondary image that was used for generating the corrugation on the model. The black lines show the location of elevated longitudinal veins in comparison with the membranes. The use of Figure 5b as a secondary input image results in the development of a model that is shown in Figure 5c,d from both dorsal and ventral sides.

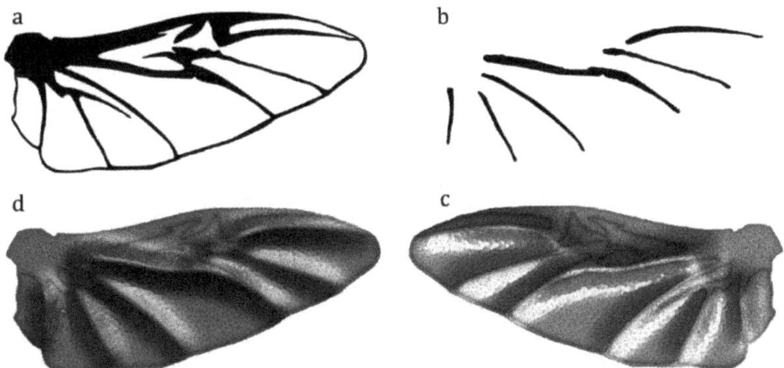

Figure 5. Modeling of the hind wing of the beetle *Allomyrina dichotoma* (Coleoptera: Scarabaeidae). (**a**) Black and white image of the wing. (**b**) The secondary image for generating corrugations showing the location of the elevated veins. (**c**) Dorsal view of the generated model. (**d**) Ventral view of the generated model.

5. Advantages of *WingMesh*

WingMesh offers several advantages over existing manual modeling techniques using commercial software packages, such as CATIA, SolidWorks, Abaqus etc.:

- The application is user-friendly and can remarkably reduce the modeling costs.
- Two-dimensional modeling using *WingMesh* is possible by the use of only an image of a given domain.
- Modeling three-dimensional (3D) out-of-plane domains is simple and can be done by the use of one additional image that contains information on corrugated spots.
- *WingMesh* can develop meshed models of domains that consist of several subdomains and discontinuities.
- *WingMesh* is particularly useful for modeling of a large number of insect wings for comparative investigations.
- Considering the use of computer vision to extract geometric wing features, *WingMesh* is applicable for insect wings that contain a high degree of geometric complexity.
- The input image for *WingMesh* should have only sufficient resolution. This is in contrast to existing tools for extracting morphological features of insect wings using an image, which usually requires high-resolution images at a large size [23,24].

WingMesh has improved the applicability of *Distmesh2D*, as listed below:

- Extracting the distance function for complex geometries is a time-consuming and error-prone task, which has been overcome by the use of the computer vision in WingMesh.
- WingMesh generates a *.inp file as the output, which is a frequently used file format.
- WingMesh has an improved ability to mesh structures that contain many discontinuities. This ability was poor in Distmesh2D, especially when dealing with domains with more than one discontinuity.
- In contrast to Distmesh2D, that can mesh domains that have no subdomains, WingMesh is capable of modeling domains with numerous subdomains.
- Compared with Distmesh2D, WingMesh can model out-of-plane domains.

6. Applications

The application presented in this study can be used for modeling a wide range of objects in both science and engineering, where a planar FE model is required. For example, models developed by our application could be used to understand the mechanical behavior of biological structures, such as insect wings, plant leaves, etc. (see [25] for more examples). In engineering, it can be employed for FE modeling of plate and shell structures used in aircraft, space crafts, ships, pressure vessels, etc. (see [26,27] for more examples). Our method could also be used in geology and geo-mechanics for the prediction of the mechanical response of complex inhomogeneous rock and concrete structures.

Although *WingMesh* is a promising first step towards the automatic modeling of insect wings, there still remain some other structural features that can be included in a wing model. A few examples of such features are nodus and vein joints, which play key roles in wing deformations both during flight [9,28,29] and at rest (i.e., wing folding [30–32]). Hence, as developers of *WingMesh*, we are currently working to develop the next generation of our program, which is able to create wing models with more structural details.

More information on *WingMesh* is available on our website: https://wingquest.org/wingmesh/.

Supplementary Materials: The following are available online at https://doi.org/10.6084/m9.figshare.12355163: Code S1: The source code of *WingMesh*. Code S2: The source code of the BA. Code S3: The GUI of *WingMesh*. Method S1: The description of the *WingMesh* code. Video S1: Finding the points inside a domain using the BA. Video S2: A tutorial about the use of *WingMesh*. File S1: The *.inp* file of Example 1. File S2: The *.inp* file of Example 2. File S3: The *.inp* file of example 3. File S4: The *.inp* file of example 4. File S5: The *.inp* file of example 5. File S6: The *.inp* file of example 6. File S7: The *.inp* file of example 7. File S8: The *.inp* file of example 8. File S9: The *.inp* file of example 9. File S9: The *.inp* file of example 9. File S10: The *.inp* file of example 10.

Author Contributions: Conceptualization: S.E., V.N., A.D., S.N.G., H.R.; Data Curation: S.E.; Funding Acquisition: S.E., S.N.G., H.R.; Investigation: H.R., S.E.; Methodology: S.E., V.N., H.R.; Project administration: S.N.G., A.D., H.R.; Resources: S.N.G., A.D.; Software: S.E., V.N.; Supervision: S.N.G., H.R., A.D.; Validation: S.E., H.R., V.N.; Visualization: S.E.; Writing—original draft preparation: H.R., S.E.; Writing—review and editing: S.E., V.N., S.N.G., A.D., H.R. All authors have read and agreed to the published version of the manuscript.

Funding: This research was funded by the German Academic Exchange Service (DAAD) to S.E., grant number 57440921.

Acknowledgments: The authors are grateful to Peyman Mayeli (Monash University, Australia) for his valuable comments at the beginning of this study. We also want to thank Zeynab Ranjbar (Ahrar Institute of Technology and Higher Education, Iran) for her assistance regarding the preparation of images.

Conflicts of Interest: The authors declare no conflict of interest. The funders had no role in the design of the study; in the collection, analyses, or interpretation of data; in the writing of the manuscript, or in the decision to publish the results.

References

1. Bathe, K.J. Finite element method. In *Wiley Encyclopedia of Computer Science and Engineering*; John Wiley & Sons, Inc.: Hoboken, NJ, USA, 2007; pp. 1–12.
2. Logan, D.L. *A First Course in the Finite Element Method*; Cengage Learning: Boston, MA, USA, 2011.

3.   Adeli, H. (Ed.) *Supercomputing in Engineering Analysis*; CRC Press: Boca Raton, FL, USA, 1991.

4.   Rao, S.S. *The Finite Element Method in Engineering*; Butterworth-Heinemann: Oxford, UK, 2017.

5.   Huebner, K.H.; Dewhirst, D.L.; Smith, D.E.; Byrom, T.G. *The Finite Element Method for Engineers*; John Wiley & Sons, Inc.: Hoboken, NJ, USA, 2001.

6.   Panagiotopoulou, O. Finite element analysis (FEA): Applying an engineering method to functional morphology in anthropology and human biology. *Ann. Hum. Boil.* **2009**, *36*, 609–623. [CrossRef] [PubMed]

7.   Dumont, E.R.; Grosse, I.R.; Slater, G.J. Requirements for comparing the performance of finite element models of biological structures. *J. Theor. Boil.* **2009**, *256*, 96–103. [CrossRef] [PubMed]

8.   Maas, S.A.; Ellis, B.J.; Ateshian, G.A.; Weiss, J.A. FEBio: Finite elements for biomechanics. *J. Biomech. Eng.* **2012**, *134*, 11005. [CrossRef] [PubMed]

9.   Rajabi, H.; Gorb, S.N. How do dragonfly wings work? A brief guide to functional roles of wing structural components. *Int. J. Odonatol.* **2020**, *23*, 23–30. [CrossRef]

10. MacNeil, J.A.; Boyd, S.K. Bone strength at the distal radius can be estimated from high-resolution peripheral quantitative computed tomography and the finite element method. *Bone* **2008**, *42*, 1203–1213. [CrossRef] [PubMed]

11. Silva, E.C.N.; Walters, M.C.; Paulino, G.H. Modeling bamboo as a functionally graded material. In *AIP Conference Proceedings*; American Institute of Physics: College Park, MD, USA, 2008; Volume 973, pp. 754–759.

12. Rajabi, H.; Jafarpour, M.; Darvizeh, A.; Dirks, J.H.; Gorb, S.N. Stiffness distribution in insect cuticle: A continuous or a discontinuous profile? *J. R. Soc. Interface* **2017**, *14*, 20170310. [CrossRef] [PubMed]

13. Toofani, A.; Eraghi, S.H.; Khorsandi, M.; Khaheshi, A.; Darvizeh, A.; Gorb, S.; Rajabi, H. Biomechanical strategies underlying the durability of a wing-to-wing coupling mechanism. *Acta Biomater.* **2020**, *110*, 188–195. [CrossRef] [PubMed]

14. Edelsbrunner, H. *Geometry and Topology for Mesh Generation*; Cambridge University Press: Cambridge, UK, 2001.

15. Jin, T.; Goo, N.S.; Park, H.C. Finite element modeling of a beetle wing. *J. Bionic Eng.* **2010**, *7*, S145–S149. [CrossRef]

16. Voo, L.; Kumaresan, S.; Pintar, F.A.; Yoganandan, N.; Sances, A. Finite-element models of the human head. *Med. Boil. Eng.* **1996**, *34*, 375–381. [CrossRef] [PubMed]

17. Cakmakci, M.; Sendur, G.K.; Durak, U. Simulation-based engineering. In *Guide to Simulation-Based Disciplines*; Springer: Cham, Switzerland, 2017; pp. 39–73.

18. Persson, P.O.; Strang, G. A simple mesh generator in Matlab. *SIAM Rev.* **2004**, *46*, 329–345. [CrossRef]

19. *Abaqus v6.7. Analysis User's Manual*; Simulia: Johnston, RI, USA, 2007.

20. Lindquist, W.B.; Lee, S.M.; Coker, D.A.; Jones, K.W.; Spanne, P. Medial axis analysis of void structure in three-dimensional tomographic images of porous media. *J. Geophys. Res. Solid Earth* **1996**, *101*, 8297–8310. [CrossRef]

21. Chopp, D.L. Some improvements of the fast marching method. *SIAM J. Sci. Comput.* **2001**, *23*, 230–244. [CrossRef]

22. Eshghi, S.; Rajabi, H.; Darvizeh, A.; Nooraeefar, V.; Shafiei, A.; Mostofi, T.M.; Monsef, M. A simple method for geometric modelling of biological structures using image processing technique. *Sci. Iran. Trans. B Mech. Eng.* **2016**, *23*, 2194–2202. [CrossRef]

23. Tofilski, A. DrawWing, a program for numerical description of insect wings. *J. Insect Sci.* **2004**. [CrossRef]

24. Mengesha, T.E.; Vallance, R.R.; Barraja, M.; Mittal, R. Parametric structural modeling of insect wings. *Bioinspir. Biomim.* **2009**, *4*, 036004. [CrossRef] [PubMed]

25. Kubínová, L.; Janáček, J.; Albrechtová, J.; Karen, P. Stereological and digital methods for estimating geometrical characteristics of biological structures using confocal microscopy. In *From Cells to Proteins: Imaging Nature across Dimensions*; Springer: Dordrecht, The Netherlands, 2005; pp. 271–321.

26. Kienzler, R.; Altenbach, H.; Ott, I. (Eds.) *Theories of Plates and Shells: Critical Review and New Applications*; Springer Science & Business Media: Berlin/Heidelberg, Germany, 2013; Volume 16.

27. Hoff, N. Thin shells in aerospace structures. In Proceedings of the 3rd Annual Meeting, Boston, MA, USA, 29 November–2 December 1966; p. 1022.

28. Rajabi, H.; Ghoroubi, N.; Stamm, K.; Appel, E.; Gorb, S.N. Dragonfly wing nodus: A one-way hinge contributing to the asymmetric wing deformation. *Acta Biomater.* **2017**, *60*, 330–338. [CrossRef] [PubMed]

29. Wootton, R.J. Functional morphology of insect wings. *Annu. Rev. Entomol.* **1992**, *37*, 113–140. [CrossRef]

30. Haas, F.; Wootton, R.J. Two basic mechanisms in insect wing folding. *Proc. R. Soc. Lond. Ser. B Biol. Sci.* **1996**, *263*, 1651–1658.

31. Haas, F.; Gorb, S.; Wootton, R.J. Elastic joints in dermapteran hind wings: Materials and wing folding. *Arthropod Struct. Dev.* **2000**, *29*, 137–146. [CrossRef]

32. Saito, K.; Nomura, S.; Yamamoto, S.; Niiyama, R.; Okabe, Y. Investigation of hindwing folding in ladybird beetles by artificial elytron transplantation and microcomputed tomography. *Proc. Natl. Acad. Sci. USA* **2017**, *114*, 5624–5628. [CrossRef] [PubMed]

Review

Nectar Feeding by a Honey Bee's Hairy Tongue: Morphology, Dynamics, and Energy-Saving Strategies

Hao Wang [1], Zhigang Wu [1], Jieliang Zhao [2] and Jianing Wu [1,*]

1   School of Aeronautics and Astronautics, Sun Yat-Sen University, Guangzhou 510006, China; wangh659@mail2.sysu.edu.cn (H.W.); wuzhigang@mail.sysu.edu.cn (Z.W.)
2   School of Mechanical Engineering, Beijing Institute of Technology, Beijing 100081, China; 6120190098@bit.edu.cn
*   Correspondence: wujn27@mail.sysu.edu.cn

Simple Summary: This paper reviews the interdisciplinary research on nectar feeding behaviour of honey bees ranging from morphology, dynamics, and energy-saving strategies, which collects a range of knowledge of feeding physiology of honey bees and may inspire the design paradigms of next-generation multifunctional microfluidic transporters.

Abstract: Most flower-visiting insects have evolved highly specialized morphological structures to facilitate nectar feeding. As a typical pollinator, the honey bee has specialized mouth parts comprised of a pair of galeae, a pair of labial palpi, and a glossa, to feed on the nectar by the feeding modes of lapping or sucking. To extensively elucidate the mechanism of a bee's feeding, we should combine the investigations from glossa morphology, feeding behaviour, and mathematical models. This paper reviews the interdisciplinary research on nectar feeding behaviour of honey bees ranging from morphology, dynamics, and energy-saving strategies, which may not only reveal the mechanism of nectar feeding by honey bees but inspire engineered facilities for microfluidic transport.

Keywords: honey bee; mouth parts anatomy; nectar feeding behaviour; dynamics; energy-saving strategies

Citation: Wang, H.; Wu, Z.; Zhao, J.; Wu, J. Nectar Feeding by a Honey Bee's Hairy Tongue: Morphology, Dynamics, and Energy-Saving Strategies. *Insects* **2021**, *12*, 762. https://doi.org/10.3390/insects12090762

Academic Editor: Johan Billen

Received: 22 June 2021
Accepted: 18 August 2021
Published: 24 August 2021

1. Introduction

The majority of flower-visiting insects, including bees, wasps [1,2], flies [3], butterflies [4], moths [5], and some beetles [6,7], obtain nutrition from floral nectar and pollen from flowering plants [8]. The honey bee (*Apis mellifera ligustica*) is a typical pollinator in the world [9]. The specialized proboscis is of great importance for a honey bee to load nectar rapidly and efficiently. The mouth parts of a honey bee are comprised of a pair of galeae, a pair of labial palpi, and a glossa [6]. The honey bee performs two feeding modes, namely lapping and suction [10]. While lapping, the honey bee drives its segmented tongue (glossa) coated in dense hairs back and forth to load nectar. When the honey bee dips nectar, the glossa protracts with the glossal hairs adhered to the glossa body. Then the glossa reaches to the maximum extension with the glossal hairs deployed. Next the brush-like glossa is filled with nectar and retracts to the mouth parts to load nectar. While sucking, the glossa extends out of the proboscis tube, directly sucking with the glossa keeping still [10].

Honey bees can feed on a range of viscous fluids at high efficiencies [8]. This behaviour is challenging because of the physical property of nectar, suggesting the nectar viscosity increases steeply with respect to the concentration, through which the glossa should have to resist high viscous drag [11–14]. In addition, if the glossa dips faster, the energetic intake rate will augment; however, the energy consumption caused by viscous drag will increase, so honey bees should have to meet the contradictive demands of both high energetic intake rate and low energetic loss while feeding on nectar. Investigations of the honey bee's feeding behaviour and related mechanical principles may reflect the health status of bees and adaptations to environmental constraints. More extensively, a healthier bee may consume less energy while feeding on nectar, who might be able to optimize the nectar harvest due to the mechanism of drag reduction. In addition, combined biological and mathematical analysis on feeding behaviour of bees may even elucidate the co-evolution between flowering plants and nectarivorous insects. In this review, we will introduce some interdisciplinary problems associated with honey bee's feeding behaviour. We will start with the anatomy of the mouth parts, followed by feeding modes, mechanism of hair erection, and energy saving strategies and conclude with potential engineering applications. The rest of the paper is structured as follows. Section 2 introduces the anatomical structure of the honey bee's mouth parts. Section 3 illustrates feeding fashion of a honey bee glossa from the perspectives of glossa kinematics and drag reduction mechanism, and Section 4 introduces the energy saving strategy by the glossa's dynamic surface. Functional compensation by regulating dipping frequency is shown in Section 5. Section 6 includes conclusions.

2. Honey Bee Mouth Parts Morphology

For the western bee (*Apis mellifera* L.), the mouth parts are comprised of a pair of galeae, a pair of labial palpi, and a hairy glossa, namely, the glossa in length of 2 mm (Figure 1a–c) [6,12]. Bushy glossal hairs in length of 100 μm, with diameter of 1~3 μm, are attached to the surface of the kidney-shaped sheath, which appear annulated on the glossa (Figure 1c–d). A thin membrane is attached to the edges of the sheath, which is also next to the corresponding sides of the muscular rod inside the glossa [15]. Notably the glossa has ~120 segments and is structured in a compliant manner. In the centre of glossa, a humour-filled cavity is formed by the sheath, muscular rod, and thin membranes. The honey bees are described to have two feeding modes, namely lapping and suction [10]. For the lapping mode, the glossa moves forward and backward with glossal hairs erecting rhythmically to load the nectar (Figure 1f). For the suction mode, the glossa stays still through the proboscis tube, and the nectar is sucked up by the cibarial pump, generating flows across the glossa surface [10].

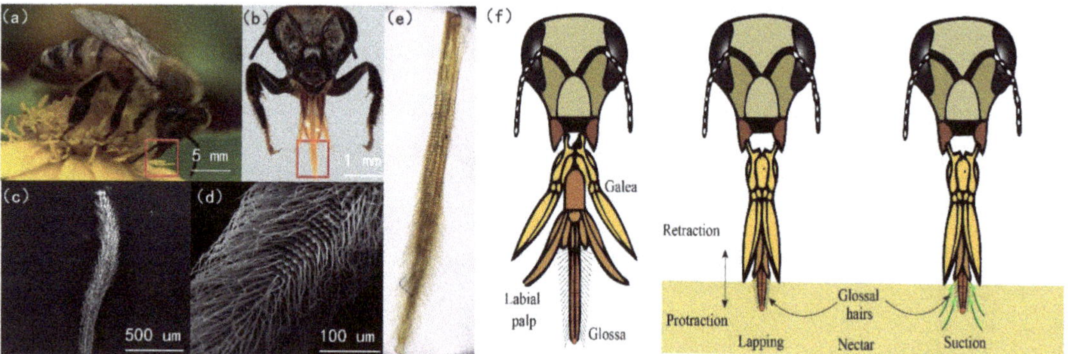

Figure 1. The honey bee's mouth parts. (**a**) A honey bee feeding on nectar on a flower. (**b**) The head and mouth parts of a honey bee. The mouth parts, highlighted in a red box, are comprised of a pair of galeae, a pair of labial palpi, and a glossa (*Apis mellifera* L.). (**c**) Scanning electron microscopic images of a bee glossa. (**d**) The glossa with bushy hairs. (**e**) The glossa observed under a microscope. (**f**) Lapping and sucking modes of a honey bee. The galeae and labial palpi form the probocid tube, then the glossa makes reciprocating movements through the tube to lap nectar [16].

3. Feeding Behaviour of a Honey Bee Glossa

3.1. Section-Wise Wettability of the Glossa

Wettability is the ability of surface to be wetted by liquid, which is determined by the balance of surface energy in the interface between air, liquid, and solid materials [17]. A honey bee propels its glossa to lap the viscous nectar, so the wettability of a honey bee glossa may be related to the nectar trapping capability [18], and the contact angle (wetting angle) is a measure of the wettability of a solid by a liquid. Generally, if the water contact angle is smaller than 90°, the solid surface is considered hydrophilic and inversely if the contact angle is bigger than 90°, the solid surface is considered hydrophobic. As a result, it is necessary to test the wettability of a honey bee glossa before we examine its feeding capability [19]. For bee flowers, the average nectar concentration in nature is 36% [20], so 25%, 35%, and 45% sucrose solutions were prepared for lab tests. Under a microscope, the contact angles measured in different glossal regions are shown in Figure 2. The results indicate that the contact angles turn smaller when using the sucrose solution with higher concentration, which insinuates that the surface exhibits stronger hydrophilicity to the thicker nectar. More extensively, the ranking of section-wise hydrophilicity suggests that the dorsal side is much easier to be wetted than the ventral side (Figure 2). We calculated the *p*-value as 0.03 between the two data sets of contact angles on the ventral surface and the dorsal surface, respectively, which suggests a significant difference between these two data sets, denoting that the dorsal surface is much easier to be wetted than the ventral surface for the reason of that the ranking of hydrophilicity, namely, D > A, E > B, and F > C. In order to better understand the comparison groups, we indicate them by stars shown in Figure 2. Moreover, the glossa tip is more hydrophilic than the middle region of the glossa, and the proximal part is the hardest to be wetted by the nectar. The section-wise wettability of the glossa may be caused by the chemical and geometrical differences on these hairy segments; the glossa surface is more hydrophilic to the higher-concentration nectar, which might be beneficial for nectar trapping, especially for the dynamic glossa surfaces. The combined chemical and geometrical differences on these hairy segments may contribute to a high flexibility in adaptation to varying environments, for instance, a broad range of liquid viscosities found in floral sources. In the next subsection, we will introduce the feeding pattern of honey bees via high-speed filming under the conditions of foraging on nectar with varying viscosities.

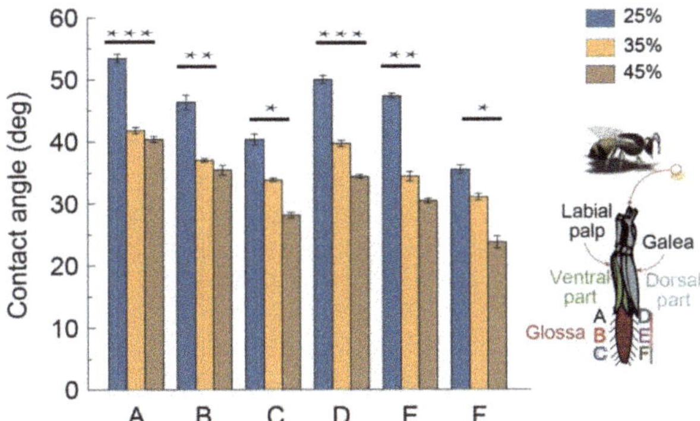

Figure 2. Section-wise wettability of a honey bee's glossa. Six regions of the glossa immersed in nectar, marked with A~F, and A~C, and D~F, represent the ventral part and dorsal part respectively. Contact angles of different regions on the glossa surface of 25%, 35%, and 45% sucrose solution. The glossa surface is more hydrophilic to the higher-concentration nectar, which is elucidated from the decreasing of contact angles with the increased nectar concentration are shown in the histogram [21]. Asterisks indicates the comparison groups.

3.2. Facultative Feeding Modes in Honey Bees

The feeding pattern of a honey bee was previously defined as "lapping", which refers to reciprocating movements of its glossa entraining nectar by the glossal hairs. However, bees would also directly suck nectar with glossa keeping protracting and staying still (Figure 3a). Wei et al. [10] demonstrate that bees have facultative feeding behaviour. Bees prefer to suck the nectar with low sugar concentration and they tend to lap the nectar with more sugar content (Figure 3b). Further lab tests showed that honey bees can switch between these two feeding patterns to choose a more efficient ingesting mode (Figure 3c). The capillary-based lapping mechanism that allows honey bees to achieve high energy intake rates when feeding on highly concentrated nectar [22], while sucking directly with glossa protracting and staying still facilitates feeding on less viscous nectar (Figure 3d,e), besides the energy intake rate E' was calculated by $E' = \rho scQ/100$, where ρ denotes the density of the nectar, s the sugar concentration, c the energy content per unit mass of sugar, and Q the nectar intake rate. Experiments validated that the key stimulus of choosing the ingesting technique is the viscosity of the nectar, rather than sugar content, according to the result that most bees feed on nectar with 10% sugar concentration, but with viscosity equivalent to 50% concentration (by adding Tylose) exhibited lapping pattern. This facultative drinking mode that is behaviourally adjusted to fluid viscosity has potentially enhanced the adaptability of honey bees to a wider range of nectar resources [23–25].

3.3. Kinematics of the Glossa and Glossal Hairs

For the lapping mode, the glossa extends out of the proboscid tube structured by the labial palpi and galeae, with glossal hairs attaching on the glossa body. Then the glossa moves back into the proboscis tube and glossal hairs flatten to offload the nectar. Here is a kinematic asymmetry, in which glossa protracts faster in a spear-like shape and retracts more slowly in a brush-like configuration. This asymmetry functions as a strategy to save energy, especially reducing the energetic consumption induced by viscous fluidic drag [26]. By observing kinematics of the glossa and glossal hairs by high-speed filming, Zhao [16] further found a specific asynchronization between glossa movements and glossal hair erection. A physical model is proposed to describe the feeding process considering the trade-off between nectar-intake volume and energy consumption. This asynchronization may be caused by the material properties of the elastic rod and the compliance of the segmented structures, especially the zig-zag shaped intersegmental membranes of the glossa [12], which is validated to be effective in maximizing the nectar-intake amount by theoretically figuring out the optimal moment when the glossal hairs begin to erect. This asynchronization suggests that the honey bee glossa can perform a scheduled coordination between glossa movements and hair erection, which could serve as valuable models for developing miniature pumps that are both extendable and have dynamic surfaces.

To uncover the anatomical mechanism of the coordination of glossa extension and hair erection, Zhu [27] compared hair erection and segment elongation and discovered a high consistency of their kinematics during the drinking process (Figure 4). In a dipping cycle, when the average erection angle of glossal hairs increases from 20 deg. to 38 deg., the average length of one glossal segment increases from 22.9 ± 1.6 µm to 24.7 ± 2.2 µm. The concordance equation was applied for evaluating correlation between these variables. The concordance measure is equal to 0.99 in the in vivo observation experiments, which shows that the average elongation of a glossal segment is closely correlated to the average erection angle of hairs.

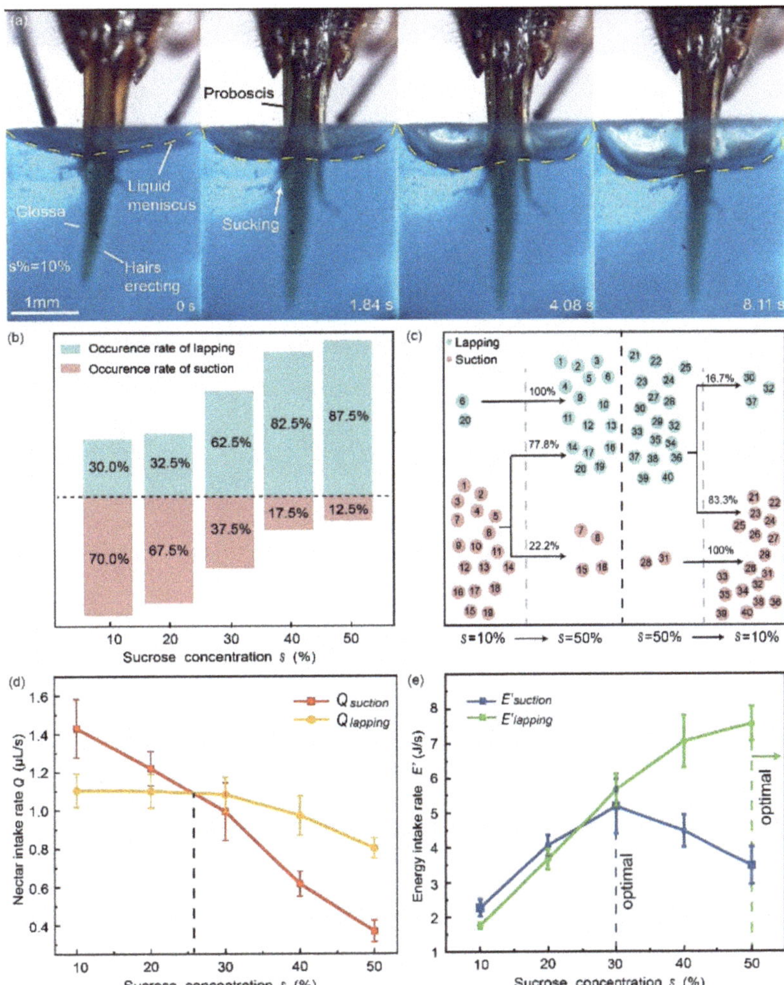

Figure 3. Switchable feeding pattern in a honey bee. (**a**) High-speed images of a honey bee sucking the artificial nectar. (**b**) Occurrence rates of the two feeding modes in honey bees, when feeding on sucrose solutions with various concentrations [10]. (**c**) Occurrence rates of switching between feeding modes when offered extreme nectar concentrations, and the dotted lines represent binary feeding mechanisms in various nectar concentrations. Each encircled number represents a different individual. (**d**) Nectar intake rates of suction and lapping under different concentrations nectar, dashed line denotes the equivalent point of feeding efficiency and the corresponding sugar concentration, the dotted line denotes an equal nectar intake rate of the feeding modes under a specific nectar concentration. (**e**) Energy intake rates of suction and lapping under different nectar concentrations. Blue dashed line depicts the optimal concentration for suction mode, and green dashed line denotes that the optimal concentration for lapping mode is around 50% or above.

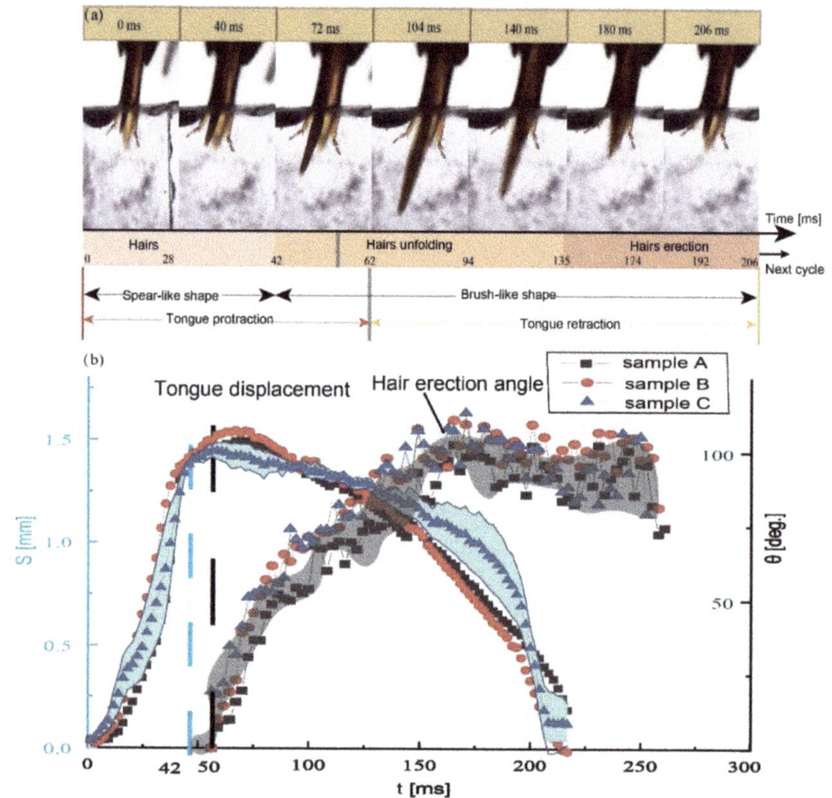

Figure 4. Morphological changes in glossal surfaces during dipping nectar and surface configurations through stretching the honey bees' glossae. (**a**) Dipping pattern of a honey bee tongue. (**b**) Asynchronization between tongue displacement and average hair erection angle. Both the in vivo and postmortem observations reveal that shortening and lengthening of the glossal segments is highly coordinated with the erection of glossal hairs, which aids in developing deformable gaps between rows of glossal hairs during nectar trapping [16].

To further demonstrate the relationship between glossa protraction and hair erection, Zhu [27] stretched the glossae of honey bees and observed them under a microscope. Glossal hairs of a dead honey bee's glossa erect only when the glossa is stretched by an external force, suggesting that the elongation of the glossal segments is coordinated with the hair erection (Figure 5). The average erection angle of hairs varies from 23 deg. to 57 deg., and the average length of one glossal segment increases from 34.7 ± 2.8 μm to 37.7 ± 3.1 μm accordingly. The concordance value is calculated as 0.96, which is close to the observations on living animals (0.99). This highly-coordinated motion indicates that the glossal hairs are hinged in the intersegmental membranes, which could deploy and fold synchronously.

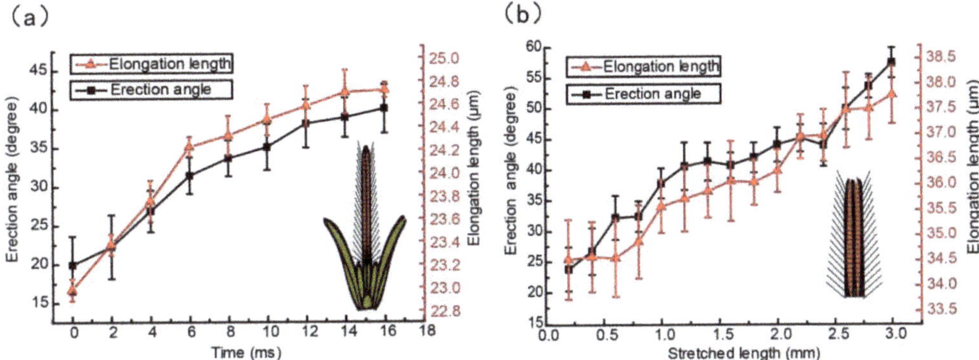

Figure 5. Coordinated movements of hair erection and segment elongation of the glossa. (**a**) Natural behaviour of nectar drinking. (**b**) Highly-coordinated movements of the glossal segments and hair erection through stretching the glossa under a microscope [27].

3.4. Coordinated Movements of the Abdomen While Dipping Nectar

As a lapper, honey bee uses a mop-like glossa to trap nectar from flowering plants. By filming the feeding honey bees, a significant increase in abdominal pumping frequency was observed when honey bees drink the sucrose solution [16]. Zhao [16] combined high-speed filming, X-ray phase contrast imaging, and mathematical models to investigate the effect of abdominal pumping in liquid feeding of honey bee. A honey bee performs abdominal pumping during feeding, which is in concordance with reciprocating movements of the glossa (Figure 6). The modelling framework demonstrates that the abdominal pumping powers the honey bee's feeding efficiency and saves foraging time. The combined experimental and theoretical investigations extend the knowledge about the function of abdominal movements, which is considered only for adjusting flight attitude or crawling through honeycombs [28]. This behaviour is functionally analogous to power suction feeding in some fish that uses most power of axial swimming muscles not only by the cranial muscles [29]. The multifunctional use of muscular actuations fulfils the switchable requirements of these animals and makes the organs structurally compacted and efficient.

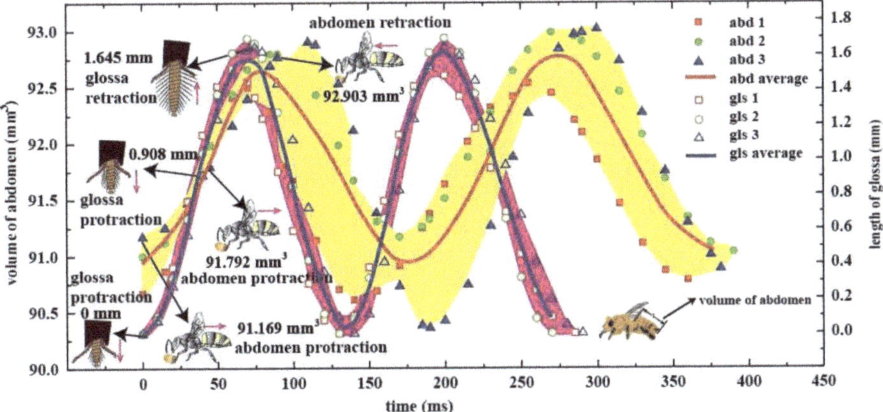

Figure 6. Dependence of abdominal and glossal movements (three samples are shown here). Both the glossa and abdomen protracted and retracted periodically, the pink arrow denotes the direction of protraction and retraction of the glossa and abdomen, thereby showing an approximate sinusoidal principle. The left vertical axis shows the volume of abdomen when lapping nectar against time, and the right vertical axis shows the real-time length of the glossa [15].

4. Energy Saving Strategy

For the viscous dipping mode, a honey bee should have to meet the combined requirement of both high energetic intake rate and low energetic dissipation caused by viscous drag. A honey bee may have to make millions of reciprocating movements during its whole life, so an energy-saving mechanism may be required to reduce the energy consumption and lower the possibility of wear caused by the viscous drag. This section includes some interdisciplinary work that covers morphology, high-speed imaging, and lubrication models, to uncover the energy saving strategy while feeding on nectar [30].

4.1. Modelling for Energy Saving

Figure 7 shows the actual glossa kinematics and hypothetical cases with various kinematic apportionment. A 7-order Fourier function that fits the actual glossa velocity in a dipping cycle ($R^2 = 0.9853$) is shown as

$$u(t) = K_1 f(t) = K_1 \left(a_0 + \sum_{i=1}^{7} (a_i \cos(\omega t) + b_i \sin(\omega t)) \right) \tag{1}$$

where $f(t)$ is the fitting equation; a_0, ω, a_i and b_i ($1 \leq i \leq 7$) are the parameters calculated by Matlab to obtain the best fit for the scatter plot; and K_1 (137 μm/cm) is a coefficient that links the sizes in high-speed photographs into those for an actual honey bee. It can be demonstrated that the honey bee can reduce its energy expenditure using the derived protraction kinematics. The power required for resisting viscous drag can be estimated as $P_v \sim \mu L u^2$, where μ is the nectar viscosity and L is the glossa length during protraction, and the power to drive glossa can be estimated as $P_t \sim mu'u \sim \rho_t a^2 u^3$, where a and ρ_t are the radius of the glossa and the density of the glossa, respectively. Since the ratio $P_t/P_v \ll 1$, the effect of P_t can be neglected [26]. The viscous drag can be written as $F_v \propto \mu \cdot (2\pi a) \cdot x(t) \cdot u(t) = K_2 \mu a \cdot x(t) \cdot u(t)$, where K_2 is a proportionality coefficient. Combining Equation (1) and the formula $x(t) = \int_0^t f(t) \mathrm{d}t$, the power needed to overcome viscous drag reads.

$$P_v(t) = F_v \cdot u(t) = \mu a K_2 K_1^3 f(t)^2 \int_0^t f(t) \mathrm{d}t \tag{2}$$

One can evaluate the benefits of keeping the glossal hairs still during tongue protraction from Equation (1). If the honey bee erects the hairs, the outer diameter of the glossa radius will augment from a to $(a + h\cos\theta)$, which will lead to a significant increase in $P_v(t)$. Scanning electron microscope imaging indicates $a \approx 50$ μm and $h \approx 170$ μm, and since $\theta \approx 45°$, we then arrive at $(a + h\cos\theta)/a = 3.4$, which means that hair erection increases the resistance by more than three times. Therefore, the honey bee is equipped with a specific glossal hair erection pattern for energy saving, where the flatten hairs can reduce viscous drag during protraction, whereas the hairs erect to trap more nectar in a single cycle during retraction. When a glossa makes reciprocating movements through viscous fluid, the viscous drag will exert on the hairy glossa surface. The nectar is a specific solution, the physical property of which is analogous to the sucrose solution, and its viscosity rises steeply with respect to concentration. From the perspective of Fluid Mechanics, the viscous drag dissipates energy so we should have to consider the energy dissipation linked to viscous drag. Some previous tests were made to validate the fact that more viscous nectar causes higher rate of wear, which indicates the viscous drag can accelerate the structural deterioration on the seemingly fragile glossal hairs.

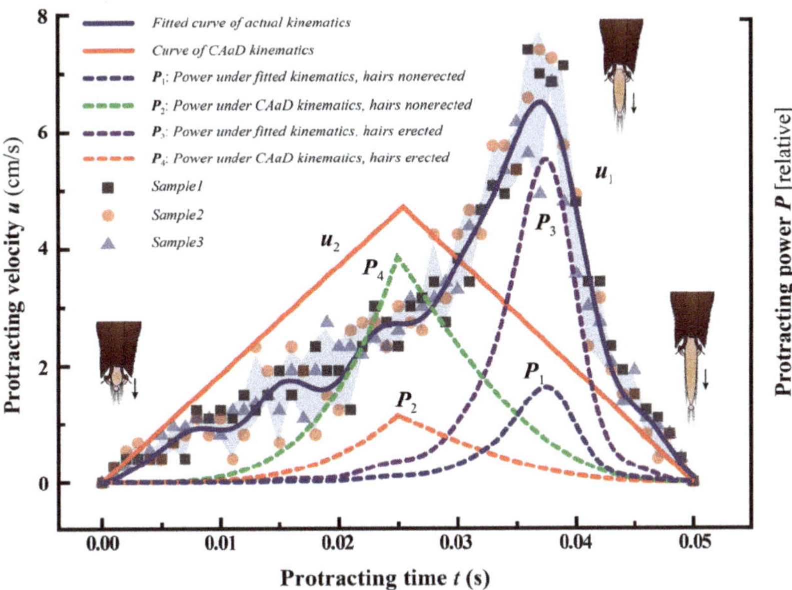

Figure 7. Energy saving by specific glossa kinematics. The scattered points show three independent protraction velocities measured from the high-speed video and the light blue area indicates the error band of the velocities. The bold blue curve represents the Fourier kinematics u_1, which fit the scatter plot well. The bold red curve shows the constant-acceleration-and-deceleration (CAaD) kinematics u_2. The dotted blue line P_1 and the dotted red line P_2 indicate the protraction power under the fitted kinematics and constant-acceleration-and-deceleration kinematics, respectively [26].

4.2. Effects of Galea Ridges on Drag Reduction

Biological surfaces with unique microstructures in nature may perform specific functions, such as impact absorption and drag reduction in dung beetles or sharks, respectively [31]. The honey bee, *Apis mellifera* L., dips viscous nectar at a high rate which is about 5 Hz by the glossa, which causes non-negligible fluidic drag that results in structural and functional deterioration. By postmortem examination, Li [32] found the ridges are parallel distributed on the inner wall of the galeae and validated its effects on drag reduction. Li then compared the structural discrepancy between workers and drones and proposed some implications about the caste-related behaviour [32].

Scanning electron microscopy (SEM) images indicate that the honey bee galea has internally transverse ridges uniformly distributed (Figure 8). Theoretical analysis show that the ridges on the galeae of honey bee's mouth parts of workers can reduce the friction coefficient by 86%. Li [32] then examined the dimensional diversities of the uniformly-distributed micro-ridges on inner walls of galeae among workers and drones of *Apis mellifera* L. The hydrodynamic model was used to calculate the friction coefficient in the mouth parts, further testing whether the sexually-dimensional variations of the micro-ridges could influence the effect on drag reduction. Theoretical estimations of the friction coefficient with respect to the dipping frequency show that the inner micro-ridges can significantly reduce friction during the feeding process of a honey bee. Li then compared effects of drag reduction regulated by the sexually-selected micro-ridges and demonstrated that the hydrodynamic coefficients of workers and drones are 0.011 ± 0.007 and 0.045 ± 0.010 respectively, which indicates that workers exhibit better capability of drag reduction in their mouth parts than that of drones. This discrepancy may have some more indications in caste-related work of honey bees. The main physiological requirement of drones is to find an airborne queen to mate and accordingly, so drones exhibit strong adaptations to forceful flying, and drones possess elaborate mating organs and powerful sense organs, such as big eyes and long antennae with many receptors for visual and olfactory orientations toward airborne queens [33]. Thus, although drones have bigger bodies, their mandibles are shorter, and their stomachs for honey storage are slimmer than those of workers [34]. Compared to drones, workers should have to fulfil a variety of tasks [35]. Workers tidy the hive, care the brood, nourish the larvae, drones, and the queen, and work for nest homeostasis [36,37]. Given these various duties, workers are equipped with well-developed hypopharyngeal and possess longer mouth parts than drones. Notably, adult drones are nourished by worker-prepared food, and their feeding ability is weaker than that of workers [34]. This experimental and theoretical combined research elucidated that the sexually-selected micro-ridges, developed inside workers and drones of honey bees' mouth parts, are structurally adapted to meet the demands of caste-related laborers of honey bees.

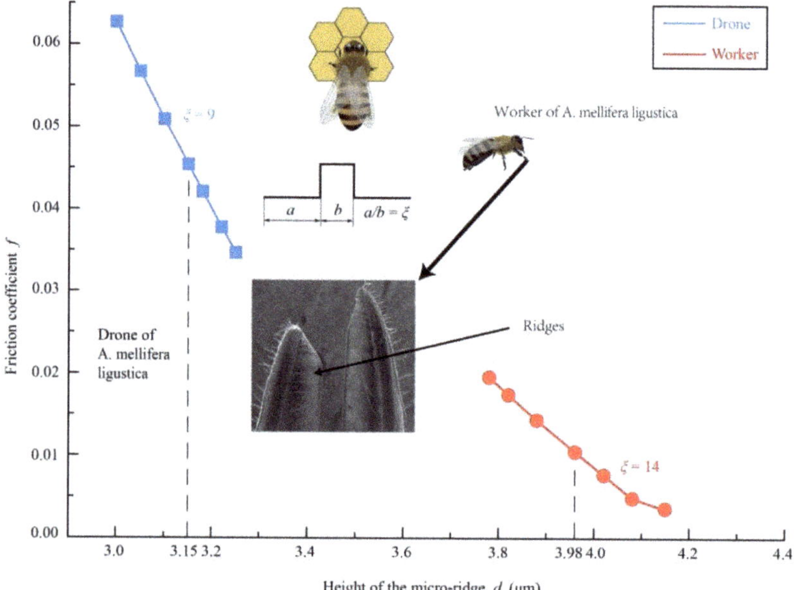

Figure 8. The friction coefficient against the heights of the micro-ridges on the inner wall of the galeae of workers and drones. The blue square and red dot represent the friction coefficient against the heights of the microridges on the inner wall of the galeae of workers and drones, respectively. Here a and b denote the dimensions of the microridges of different castes of honey bees, in which a is the length of the galea and b is the ridges, and ξ denotes the average dimensions of the workers and drones. The dotted lines illustrate the measured mean height of the microridges on the galeae of the workers was 3.98 µm, whereas that of the drones was 3.15 µm [32].

5. Functional Compensation by Regulating Dipping Frequency

Because of the highly-intensive viscous drag exerting on the glossa during nectar feeding, the glossal setae tend to wear out in the high-viscosity nectar. However, bees at varying day ages can maintain the nectar intake rate at 0.39 ± 0.03 µg·s^{-1} (35% nectar). Shi found that the average glossal setae length decreases with respect to age from 17 to 25 days, and it degrades even faster when fed with higher-viscosity nectar. Lab tests indicated that the older honey bees with short setae dip nectar more quickly. Moreover, a correlation between dipping frequency f and the average glossal setae length h, is found as $h = -15.435f + 212.04$. Based on the glossa anatomy, a fluid transport model is proposed to calculate the nectar intake rate. Theoretical analysis showed that a honey bee with shorter setae can compensate the nectar intake rate by increasing the dipping frequency. Considering the wear of the setae and dipping compensation, Shi arrived at the results that the total energy intake rate is about 106 times the power required to overcome viscous drag; the energy dissipation caused by viscous drag is negligible [26]. Therefore, the effect of augmentation of viscous drag caused by the increase of the dipping frequency on the energy intake rate of bees is almost negligible. Natural selection tends to feed quickly and efficiently, as honey bees are threatened by predators and economic necessities [38]. Therefore, honey bees must meet the contradictive demands of keeping the visit time short and the optimal nectar mass intake rate. Although the natural wear of glossal setae will affect the nectar intake rate, by adjusting the dipping frequency, both requirements can be satisfied, which is in accordance with the results from lab tests of wearing bee tongues in the 35% and 45% sucrose solutions, respectively (Figure 9).

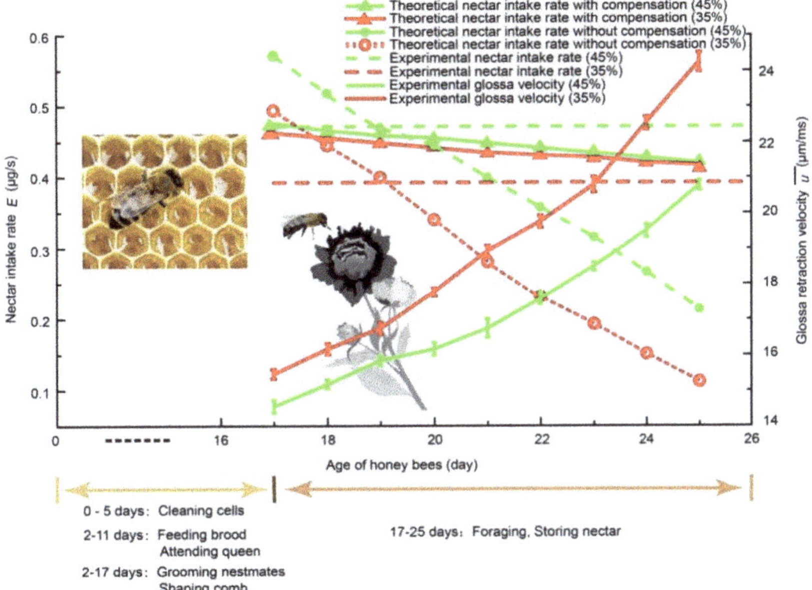

Figure 9. The honey bee augments dipping frequency to compensate for glossa hair deterioration. The relationship between theoretical nectar mass intake rate $\dot{M}$ and setae wear agrees with the experimental data captured from lab tests for dipping both the 35% and 45% sucrose solutions [39].

6. Conclusions

Investigations of feeding techniques by a honey bee's glossa are interdisciplinary work that covers morphology, behavioural dynamics, and energy-saving strategies. Future work might be extended to the following aspects. (1) The nectar property may drive the feeding property more complicated, in which, for instance, the nectar viscosity increases steeply with respect to the nectar concentration, and it is also influenced by the temperature [13]. The flowers may have an internal microclimate which is up to 4 °C higher than the external temperature, which not only provides more heat to sustain the thoracic temperature of honey bees especially in winter time but makes the nectar a bit thinner, which is much easier to be digested because of the lower nectar viscosity [13]. The combined experimental and theoretical methodology is required to uncover this. (2) The dipping behaviour may be an indicator to reflect the health state of the honey bees. Air pollution, pesticide abuse, and climate change may strongly influence the honey production rate and even survival rate of the bee colony [40,41]. The dipping frequency is closely related to the energetic intake rate, so we may use the dipping frequency as a measure to evaluate the health status of the bee colony. (3) The bee cannot only feed on nectar in different floral structures but can lick dry sugar during droughts. The functional flexibility in feeding remains unexplored. The bee glossa is comprised of segmented structure which can perform a million times of reciprocating movements. How the bee glossa meets the contradictive demands of high deformability and stiffness is still unknown. Combining various experiments and theoretical frameworks, more extensive research will be conducted, not only to reveal the behavioural characteristics of honey bees but for inspiring the next-generation facilities like micropumps and other viscous fluidic transport facilities.

Author Contributions: Conceptualization, J.W. and Z.W.; resources, J.Z. and H.W.; data curation, H.W.; writing—original draft preparation, H.W. and J.W; writing—review and editing, H.W., and J.W.; supervision, Z.W. and J.W.; project administration, J.W.; funding acquisition, J.W. All authors have read and agreed to the published version of the manuscript.

Funding: This work was supported by the National Natural Science Foundation of China (grant no. 51905556), the Grant for Popularization of Scientific and Technological Innovation of Guangdong Province (grant no.2020A1414040007), and the research grant of Sun Yat-Sen University for Bairen Plan (grant no.76200-18841223).

Institutional Review Board Statement: Not applicable.

Data Availability Statement: Data is contained within the article or supplementary material. The data presented in this study are available in "Nectar feeding by a honey bee's hairy tongue: morphology, dynamics, and energy-saving strategies".

Acknowledgments: We appreciate Wei Zhang and Yu Sun from Sun Yat-Sen University for the preparation of figures.

Conflicts of Interest: The authors declare no conflict of interest.

References

1. Cook, J.M.; Rasplus, J.-Y. Mutualists with attitude: Coevolving fig wasps and figs. *Trends Ecol. Evol.* **2003**, *18*, 241–248. [CrossRef]
2. Zhang, H.; Song, J.; Zhao, H.; Li, M.; Han, W. Predicting the Distribution of the Invasive Species *Leptocybe invasa*: Combining MaxEnt and Geodetector Models. *Insects* **2021**, *12*, 92. [CrossRef] [PubMed]
3. Elberling, H.; Olesen, J.M. The structure of a high latitude plant-flower visitor system: The dominance of flies. *Ecography* **1999**, *22*, 314–323. [CrossRef]
4. Ghazanfar, M.; Malik, M.F.; Hussain, M.; Iqbal, R.; Younas, M. Butterflies and their contribution in ecosystem: A review. *J. Entomol. Zool. Stud.* **2016**, *4*, 115–118.
5. Cunningham, J.P.; Moore, C.J.; Zalucki, M.; West, S. Learning, odour preference and flower foraging in moths. *J. Exp. Biol.* **2004**, *207*, 87–94. [CrossRef]
6. Krenn, H.; Plant, J.D.; Szucsich, N. Mouthparts of flower-visiting insects. *Arthropod Struct. Dev.* **2005**, *34*, 1–40. [CrossRef]
7. Lechantre, A.; Michez, D.; Damman, P. Collection of nectar by bumblebees: How the physics of fluid demonstrates the prominent role of the tongue's morphology. *Soft Matter* **2019**, *15*, 6392–6399. [CrossRef] [PubMed]
8. Lechantre, A.; Draux, A.; Hua, H.-A.B.; Michez, D.; Damman, P.; Brau, F. Essential role of papillae flexibility in nectar capture by bees. *Proc. Natl. Acad. Sci. USA* **2021**, *118*, e2025513118. [CrossRef]
9. Breeze, T.; Bailey, A.; Balcombe, K.; Potts, S. Pollination services in the UK: How important are honeybees? *Agric. Ecosyst. Environ.* **2011**, *142*, 137–143. [CrossRef]
10. Wei, J.; Huo, Z.; Gorb, S.N.; Rico-Guevara, A.; Wu, Z.; Wu, J. Sucking or lapping: Facultative feeding mechanisms in honeybees (*Apis mellifera*). *Biol. Lett.* **2020**, *16*, 20200449. [CrossRef]
11. Kingsolver, J.G.; Daniel, T.L. Mechanical determinants of nectar feeding strategy in hummingbirds: Energetics, tongue morphology, and licking behavior. *Oecologia* **1983**, *60*, 214–226. [CrossRef] [PubMed]
12. Snodgrass, R.E.; Morse, R.A. *Anatomy of the Honey Bee*; Cornell University Press: Ithaca, NY, USA, 1985.
13. Nicolson, S.W.; Nepi, M.; Pacini, E. *Nectaries and Nectar*; Springer: Berlin/Heidelberg, Germany, 2007; Volume 4.
14. Burrows, M.; Sutton, G. Interacting Gears Synchronize Propulsive Leg Movements in a Jumping Insect. *Science* **2013**, *341*, 1254–1256. [CrossRef]
15. Zhao, J.; Meng, F.; Yan, S.; Wu, J.; Liang, Y.; Zhang, Y. Abdominal pumping involvement in the liquid feeding of honeybee. *J. Insect Physiol.* **2019**, *112*, 109–116. [CrossRef]
16. Zhao, C.; Wu, J.; Yan, S. Observations and temporal model of a honeybee's hairy tongue in microfluid transport. *J. Appl. Phys.* **2015**, *118*, 194701. [CrossRef]
17. Kumar, G.; Prabhu, K.N. Review of non-reactive and reactive wetting of liquids on surfaces. *Adv. Colloid Interface Sci.* **2007**, *133*, 61–89. [CrossRef]

18. Lehnert, M.S.; Bennett, A.; Reiter, K.; Gerard, P.D.; Wei, Q.-H.; Byler, M.; Yan, H.; Lee, W.-K. Mouthpart conduit sizes of fluid-feeding insects determine the ability to feed from pores. *Proc. R. Soc. B Biol. Sci.* **2017**, *284*, 20162026. [CrossRef] [PubMed]
19. Rico-Guevara, A.; Rubega, M.A.; Hurme, K.J.; Dudley, R. Shifting Paradigms in the Mechanics of Nectar Extraction and Hummingbird Bill Morphology. *Integr. Org. Biol.* **2019**, *1*, oby006. [CrossRef] [PubMed]
20. Pyke, G.; Waser, N.M. The Production of Dilute Nectars by Hummingbird and Honeyeater Flowers. *Biotropica* **1981**, *13*, 260. [CrossRef]
21. Wu, J.; Zhu, R.; Yan, S.; Yang, Y. Erection pattern and section-wise wettability of honeybee glossal hairs in nectar feeding. *J. Exp. Biol.* **2015**, *218*, 664–667. [CrossRef]
22. Nicolson, S.W.; De Veer, L.; Köhler, A.; Pirk, C. Honeybees prefer warmer nectar and less viscous nectar, regardless of sugar concentration. *Proc. R. Soc. B Biol. Sci.* **2013**, *280*, 20131597. [CrossRef]
23. Seeley, T.D. Social foraging by honeybees: How colonies allocate foragers among patches of flowers. *Behav. Ecol. Sociobiol.* **1986**, *19*, 343–354. [CrossRef]
24. Kakutani, T.; Inoue, T.; Kato, M. Nectar secretion pattern of the dish-shaped flower, Cayratia japonica (Vitaceae), and nectar utilization patterns by insect visitors. *Popul. Ecol.* **1989**, *31*, 381–400. [CrossRef]
25. Hunt, G.J.; Page, R.E.; Fondrk, M.K.; Dullum, C.J.; Hunt, G.J.; Page, R.E., Jr.; Fondrk, M.K.; Dullum, C.J. Major quantitative trait loci affecting honey bee foraging behavior. *Genetics* **1996**, *141*, 1537–1545. [CrossRef]
26. Wu, J.; Yang, H.; Yan, S. Energy saving strategies of honeybees in dipping nectar. *Sci. Rep.* **2015**, *5*, 15002. [CrossRef]
27. Zhu, R.; Lv, H.; Liu, T.; Yang, Y.; Wu, J.; Yan, S. Feeding Kinematics and Nectar Intake of the Honey Bee Tongue. *J. Insect Behav.* **2016**, *29*, 325–339. [CrossRef]
28. Finlayson, L.H.; Lowenstein, O. The structure and function of abdominal stretch receptors in insects. *Proc. R. Soc. Lond. Ser. B Biol. Sci.* **1958**, *148*, 433–449. [CrossRef]
29. Camp, A.L.; Roberts, T.J.; Brainerd, E.L. Swimming muscles power suction feeding in largemouth bass. *Proc. Natl. Acad. Sci. USA* **2015**, *112*, 8690–8695. [CrossRef] [PubMed]
30. Kim, W.; Bush, J.W.M. Natural drinking strategies. *J. Fluid Mech.* **2012**, *705*, 7–25. [CrossRef]
31. Hou, Q.; Yang, X.; Cheng, J.; Wang, S.; Duan, D.; Xiao, J.; Li, W. Optimization of Performance Parameters and Mechanism of Bionic Texture on Friction Surface. *Coatings* **2020**, *10*, 171. [CrossRef]
32. Li, C.; Wu, J.-N.; Yang, Y.-Q.; Zhu, R.-G.; Yan, S.-Z. Drag reduction effects facilitated by microridges inside the mouthparts of honeybee workers and drones. *J. Theor. Biol.* **2016**, *389*, 1–10. [CrossRef] [PubMed]
33. Seidl, R. Die Sehfelder und Ommatidien-Divergenzwinkel der drei Kasten der Honigbiene (*Apis mellifica*). *Verh. Zool. Gesell.* **1980**, 359–373.
34. Hrassnigg, N.; Crailsheim, K. Differences in drone and worker physiology in honeybees (*Apis mellifera*). *Apidologie* **2005**, *36*, 255–277. [CrossRef]
35. Tautz, J. *The Buzz about Bees: Biology of a Superorganism*; Springer Science & Business Media: Berlin, Germany, 2008.
36. Schmickl, T.; Crailsheim, K. How honeybees (*Apis mellifera* L.) change their broodcare behaviour in response to non-foraging conditions and poor pollen conditions. *Behav. Ecol. Sociobiol.* **2002**, *51*, 415–425.
37. Schmickl, T.; Crailsheim, K. Inner nest homeostasis in a changing environment with special emphasis on honey bee brood nursing and pollen supply. *Apidologie* **2004**, *35*, 249–263. [CrossRef]
38. Michener, C.D. *The Bees of the World*; JHU Press: Baltimore, MD, USA, 2000; Volume 1.
39. Wu, J.; Chen, Y.; Li, C.; Lehnert, M.S.; Yang, Y.; Yan, S. A quick tongue: Older honey bees dip nectar faster to compensate for mouthpart structure deterioration. *J. Exp. Biol.* **2019**, *222*, jeb212191. [CrossRef] [PubMed]
40. Le Conte, Y.; Navajas, M. Climate change: Impact on honey bee populations and diseases. *Revue Sci. Tech. Office Int. Epizoot.* **2008**, *27*, 499–510.
41. Brown, M.J.F.; Paxton, R. The conservation of bees: A global perspective. *Apidologie* **2009**, *40*, 410–416. [CrossRef]

Article

How Does the Intricate Mouthpart Apparatus Coordinate for Feeding in the Hemimetabolous Insect Pest *Erthesina fullo*?

Yan Wang and Wu Dai *

Key Laboratory of Plant Protection Resources and Pest Management of the Ministry of Education, College of Plant Protection, Northwest A&F University, Yangling 712100, China; wangyan105422@163.com
* Correspondence: daiwu@nwsuaf.edu.cn; Tel.: +89-29-8708-2098

Received: 9 July 2020; Accepted: 2 August 2020; Published: 4 August 2020

Simple Summary: To better understand the feeding mechanism of *Erthesina fullo*, the fine structure of the mouthparts is examined with scanning electron microscopy, and feeding performance are observed directly under laboratory conditions for the first time. The adult feeding process involves several steps, including exploring and puncturing of the host plant epidermis, a probing phase, an engorgement phase, and removal of the mouthparts from the host tissue. Proceeding from labium towards the mandibular stylets, the movement pattern becomes increasingly stereotypical, including the sensilla on the tip of the labium probing, the labium making an elbow-like bend between the first and second segment, the base of the stylet fascicle housing in the groove of the labrum, the mandibular stylets penetrating the site and maxillary stylets feeding. The morphology of mouthparts is similar to those of other Heteroptera. The four-segmented labium has eleven types of sensilla. The mandibular stylet tips have two nodules preapically on the convex external surface. The structure and function of the mouthparts are adapted for the phytophagous feeding habit in this species. This study increases the available detailed morphological and behavioral data for Hemiptera and will potentially contribute to improving our understanding of this pest's feeding behavior and sensory mechanisms.

Abstract: The yellow marmorated stink bug, *Erthesina fullo* (Thunberg, 1783), is a major pest of certain tree fruits in Northeast Asia. To better understand the feeding mechanism of *E. fullo*, the fine structure of the mouthparts, including the distribution and abundance of sensilla, are examined with scanning electron microscopy (SEM), and their functions are observed directly under laboratory conditions. The feeding performance is described in detail and illustrated for the first time. The adult feeding process involves several steps, including exploring and puncturing of the host plant epidermis, a probing phase, an engorgement phase, and removal of the mouthparts from the host tissue. Proceeding from labium towards the mandibular stylets, the movement pattern becomes increasingly stereotypical, including the sensilla on the tip of the labium probing, the labium making an elbow-like bend between the first and second segment, the base of the stylet fascicle housing in the groove of the labrum, the mandibular stylets penetrating the site and maxillary stylets feeding. In terms of morphology, the mouthparts are similar to those of other Heteroptera, consisting of a triangular pyramidal labrum, a tube-like and segmented labium with a deep groove on the anterior side, and a stylet fascicle consisting of two mandibular and two maxillary stylets. The four-segmented labium has five types of sensilla basiconica, three types of sensilla trichodea, two types of sensilla campaniformia and 1 type of sensilla coeloconica. Among them, sensilla trichodea one and sensilla basiconica one are most abundant. The tripartite apex of the labium is covered with abundant sensilla trichodea three and a few sensilla basiconica 5. The mandibular stylet tips have two nodules preapically on the dorsal margin of the convex external surface, which may help in penetrating plant tissue and anchoring the mouthparts. The externally smooth maxillary stylets interlock to form a larger food canal and a smaller salivary canal. The structure and function of the mouthparts are adapted for the phytophagous feeding habit in this species. Similarities and differences between the mouthparts of *E. fullo* and those of other Heteroptera are discussed.

Keywords: *Erthesina fullo*; mouthparts; sensillum; ultramorphology; feeding performance

1. Introduction

Hemiptera is the largest and most diverse non-holometabolous insect order, containing over 75,000 species. They are characterized by specialized piercing-sucking mouthparts, in which the modified mandibles and maxillae form two pairs of stylets sheathed within a modified labium [1,2]. These mouthparts facilitate feeding on fluids of various animal and plant hosts and have sensory organs used in both host location and feeding. The Hemiptera have been classified into four major taxa (suborders: Auchenorrhyncha, Sternorrhyncha, Coleorrhyncha and Heteroptera). Abundant data are available on some aspects of hemipteran mouthpart morphology based on light and scanning electron microscopy, for a few species of Auchenorrhyncha, e.g., Fulgoroidea [3,4], Cicadellidae [5–9], Aphidoidea [10–13], Coccoidea [14] and Aleyrodidae [15,16] of Sternorrhyncha and the Heteroptera [17–20]. These provide insights into feeding mechanisms and contribute to assessment of phylogenetic relationships [7,8,11,13,19,21–25]. However, so far, mouthpart morphology of some major groups remains little studied.

As a biologically successful group of organisms, the heteropterans (true bugs) are prolific and diverse and have acquired a variety of feeding habits. Some heteropterans suck surface fluids (e.g., nectar), some pierce tissues to suck sap or blood, and others obtain nourishment from dried seeds. Numerous modifications of mouthpart structures reflect the diversity of food sources and feeding habits of this group. Cobben [17] studied the heteropteran feeding stylets of 57 families and 145 species, but provided little information on the labium and the types and distributions of sensilla present on this stucture. In the carnivorous heteropterans, different feeding mechanisms are reflected in differences in the labial tip sensilla [26] and the movement and penetration of the stylets during feeding [27].

The strategies used by various phytophagous Heteroptera to feed on a variety of plant structures may include stylet-sheath feeding, lacerate and flush feeding, macerate and flush feeding, and osmotic pump feeding [17,28]. In general, feeding damage from heteropterans can be classified into five categories: Localized wilting and necrosis, abscission of fruiting forms, morphological deformation of fruits and seeds, modified vegetative growth, and tissue malformation [29–31]. All plant-sucking heteropterans are potential vectors of plant disease, and the lesions left behind at the feeding site can facilitate secondary infections by plant pathogens [32].

Pentatomidae (stink bugs) is one of the largest families within the Heteroptera. Stink bugs feed by inserting their stylets into the food source to suck up nutrients and may transmit plant pathogens, resulting in plant wilt and, in many cases, abortion of fruits and seeds. Compared with more specialized Hemiptera, Pentatomidae use diverse feeding strategies that allow them to feed from a wide range of plant structures including vegetative structures, such as stems and leaves, and reproductive plant structures such as seeds, nuts, pods and fruits [33]. Stink bug feeding can damage crops in different ways dependent upon the plant structure(s) attacked, e.g., vegetative or reproductive. Previous studies of mouthparts in Pentatomidae have mostly focused upon differences in certain aspects of the mouthparts among stinkbug species, such as the types and distribution of labial sensilla [34–36] and the internal structure of mandibular and maxillary stylets [37]. Information on the distribution of sensilla on the mouthparts and the relationships between mouthpart structure and function in feeding, useful in the classification of stink bugs, is not yet available.

The yellow marmorated stink bug *Erthesina fullo* (Thunberg, 1783) is one of the most widely distributed phytophagous pests in East Asia. It causes severe loss to many horticultural crops, such as apples, cherries and pears [38–40], and disturbs humans by invading houses in large numbers when overwintering. Both nymphs and adults of *E. fullo* primarily suck the sap from the trunk, leaves, immature stems and fruits of plants. Previous research on this species has mainly been limited to basic biology, behavior and integrated control [39,41,42]. Although, feeding damage from *E. fullo* has been characterized in agronomic crops, tree fruits and vegetables, little is known about the fine structure of the mouthparts and the feeding mechanism of *E. fullo*, and in particular, how the sensilla are distributed on the mouthparts and function in locating host plants.

Here, we used scanning electron microscopy to investigate the mouthpart morphology and distribution of sensilla of *E. fullo*. We also observed feeding behavior. The outcome of this study increases the available detailed morphological and behavioral data for Hemiptera and will potentially contribute to improving our understanding of this pest's feeding behavior and sensory mechanisms. This study provides more data for future comparative morphological studies in Pentatomidae.

2. Material

2.1. Insect Collecting

Adults of *E. fullo* used for SEM in this study were obtained from the campus of Northwest A&F University in Yangling, Shaanxi Province, China (34°16′ N, 108°07′ E, elev. 563 m) in August 2016, and were preserved in 70% ethanol and stored at 4 °C. For observing feeding behavior, additional adults of *E. fullo* were collected at the same locality in September 2019.

2.2. Samples for SEM

Ten females and twelve male specimens of *E. fullo* were fixed in 70% ethanol. The labium and the stylet, dissected using fine dissecting needles under 40× magnification (Nikon SMZ 1500, stereomicroscope, Tokyo, Japan), were prepared as study samples. The samples were cleaned in an ultrasonic bath (250 W) (KQ118, Kunshan, China) for 10 to 15 s in 70% ethanol three times, then dehydrated in serial baths of 80%, 90% and 100% ethanol each for 15 min. Samples then underwent dehydration in a mixture of 100% ethanol and 100% tert-Butanol at the ratios 3:1, 1:1, and 1:3 (by volume) for 15 min at each concentration followed by a final replacement treatment in 100% tert-Butanol for 30 min. Specimens were then freeze-dried with liquid CO_2, mounted on aluminum stubs with double-sided copper sticky tape and sputtered with gold/palladium (40/60) in a LADD SC-502 (Vermont, USA) high resolution sputter coater. The samples were subsequently examined with a Hitachi S-3400N SEM (Hitachi, Tokyo, Japan) operated at 15 kV [8] or Nova Nano SEM-450 (FEI, Hillsboro, OR, USA) at 5–10 kV.

2.3. Feeding Behavior on Different Types of Substrates

To observe the feeding behavior of the insects on fresh fruit and twigs, some fruit of orange, pear and grape as well as twigs of pear and grape were offered to twenty male and female individuals of *E. fullo* in an optical quality colorless glass enclosure 100 mm in diameter and 135 mm tall. The insects were observed intermittently throughout the feeding period for one week. Sequential images of adult feeding performance were taken using a mobile phone (vivo Y18L) with an 8-megapixel camera when conditions were suitable. The images were saved directly to a computer for later analysis.

2.4. Image Processing and Terminology

Photographs and SEMs of mouthparts were observed and measured after being imported into Adobe Photoshop CC 2019 (Adobe Systems, San Jose, CA, USA). Measurements are given as means ± standard error of the mean. Schematic diagrams were drawn with Microsoft Office Word 2007 and processed with Photoshop CC 2019 (Adobe Systems, San Jose, CA, USA). For classification of sensilla, the systems of Altner and Prillinger [43] were used in addition to the more specialized nomenclature from other studies [44].

2.5. Data Analysis

The lengths of the mouthparts were compared between sexes using a Student t-test. Statistical analyses were executed using SPSS 19.0 (SPSS, Chicago, IL, USA).

3. Results

3.1. General Morphology and Structure of Mouthparts

The mouthparts of *E. fullo* are similar to those of other heteropterans, arising from the anteroventral part of the head capsule and composed of a long labrum, a tube-like labium and a stylet bundle comprising two maxillary stylets (Mx) and two mandibular stylets (Md). The four-segmented labium has a long internal labial groove (Lg) that surrounds the stylet fascile (Sf) and is covered with different types of sensilla symmetrically distributed on the surface of both sides of the groove or on the distal surface (Figure 1A–C). The two inner maxillary stylets interlock to form the food and salivary canals; they are partially surrounded by two serrate-edged mandibular stylets. The stylet fascicle is housed inside the labial groove and proximally covered by the small cone-shaped labrum. No obvious differences were noted between the mouthpart structures of females and males except for the length of the labium (t(9) = 9.473, p = 0.000) (Table 1).

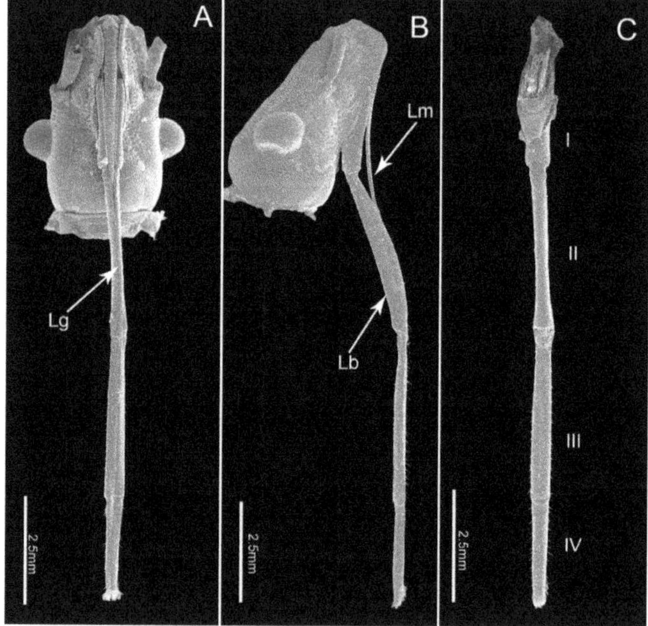

Figure 1. Scanning electron micrographs of the head of *E fullo*. (**A**) Ventral view; (**B**) Lateral view; (**C**) Dorsal view showing four-segmented labium (I–IV); Lg, labial groove; Lm, labrum; Lb, labium.

3.1.1. Labrum

The cone-shaped labrum (Lm) is attached to the anterior margin of the anteclypeus and protrudes forward beyond 2/5 length of the second segment (Figures 1A and 2A). It is closely adpressed over the first labial segment and partly embedded in the labial groove. The surface of the labrum is plicated and densely covered with regular transverse wrinkles (Figure 2A). The ventral side of the labrum bears a pair of sensilla basiconica 1 (Sb1). Sensilla trichodea 1 (St1), sensilla coeloconica (Sco) and cuticular pores (Cpo) are arranged irregularly on the ventral region of the labrum (Figure 2A–H).

Table 1. Measurements of labrum and labium (mean ± SE) obtained from scanning electron microscopy. N = sample size. Lm, labrum; Lb, labium; Lb1, first segment of labium; Lb2, second segment of labium; Lb3, third segment of labium; Lb4, fourth segment of labium.

Sex	Position	Length (µm)	Width (µm)	N
Male	Lm	4687.3 ± 310.1		6
	Lb	11503.3 ± 123.9		6
	Lb1	2013.0 ± 43.4	586.7 ± 7.9	6
	Lb2	3946.8 ± 70.7	282.2 ± 5.3	6
	Lb3	3491.6 ± 118.0	384.4 ± 5.7	6
	Lb4	2366.7 ± 59.1	290.2 ± 7.6	6
Female	Lb	12952.5 ± 75.5		5
	Lb1	2359.3 ± 22.6	629.0 ± 15.5	5
	Lb2	3771.6 ± 124.4	286.3 ± 5.3	5
	Lb3	4055.9 ± 55.0	410.0 ± 6.0	5
	Lb4	2579.4 ± 19.5	315.0 ± 3.5	5

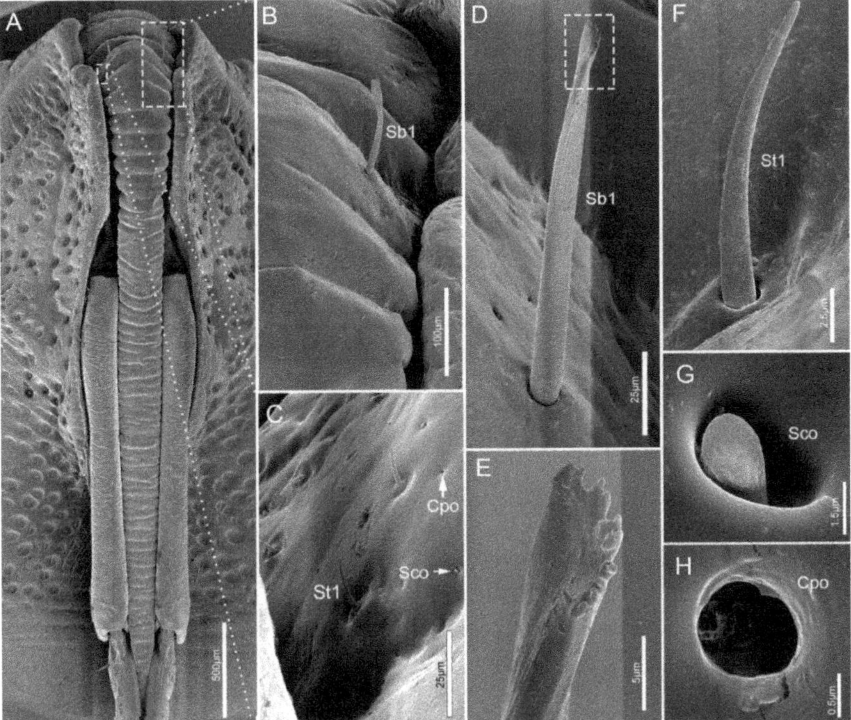

Figure 2. SEM of labrum of *E. fullo*. (**A**) Ventral view; (**B**) Enlarged view of box in (**A**), showing sensillum basiconicum 1 (Sb1); (**C**) Enlarged view of box in (**A**), showing sensilla trichodea 1 (St1), sensilla coeloconica (Sco) and cuticular pores (Cpo); (**D**) Sensillum basiconicum 1 (Sb1); (**E**) Enlarged view of box in (**D**); (**F**) Sensillum trichodeum 1 (St1); (**G**) Sensillum coeloconicum (Sco); (**H**) Enlarged view of cuticular pore (Cpo).

3.1.2. Labium

The labium, suspended from the front of the head, is tubular in shape and subdivided into four-segments externally (Figure 1A–C). The anterior surface of the labium is bisected by a deep longitudinal groove, which encases the mandibular and maxillary stylets. All segments of the labium are covered with different types of sensilla mainly positioned symmetrically on each side of the labial groove (Lg) and distally, with fewer sensilla on the posterior and lateral surfaces.

The four segments vary in size (Table 1) and morphology (Figure 1A–C). Overall the labium is broad and of uniform width through most of its length with the distal segment widening near the tip.

The proximal labial segment (Lb1), the shortest and widest of the four segments, is broad at the base, gradually narrows at the middle, and then slightly widens to the apex in posterior view (Figures 2A and 3A,B). The distal part of the dorsum is contracted inward, is crescent-shaped and no sensilla are observed on this surface (Figure 3A). Four types of sensilla (sensilla basiconica 1, sensilla trichodea 1, sensilla coeloconica, sensilla campaniformia 1) and cuticular pores (Cpo) are arranged on the ventral surface (Figure 3B–F).

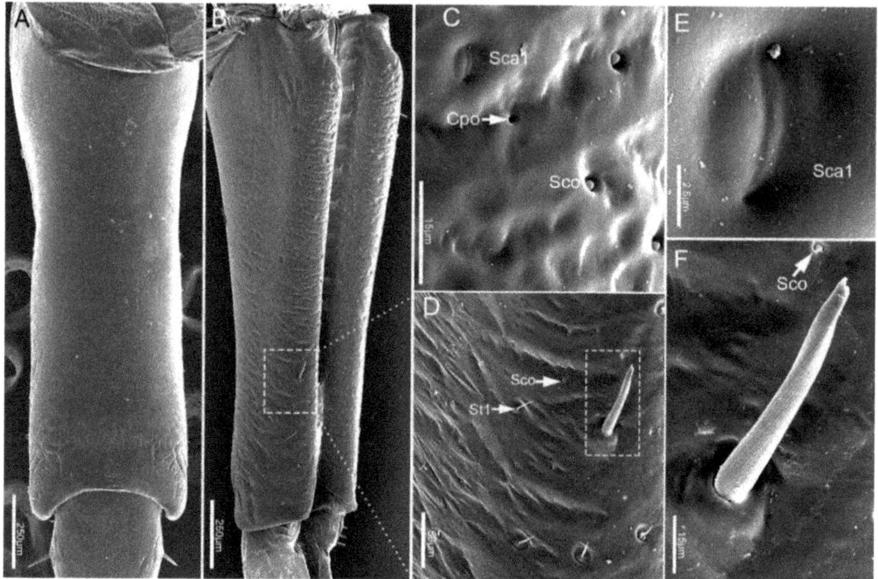

Figure 3. SEM of first labial segment of *E. fullo*. (**A**) Ventral view; (**B**) Lateral view; (**C**) Enlarged view of surface of the first segment, showing sensilla campaniformia 1 (Sca1), sensilla coeloconica (Sco) and cuticular pores (Cpo); (**D**) Enlarged view of outlined box in (**B**), showing sensilla trichodea 1 (St1), sensilla coeloconica (Sco) and sensilla basiconica 1 (Sb1); (**E**) Sensillum campaniformium 1 (Sca1); (**F**) Enlargement of outlined box in (**D**), showing sensillum coeloconicum (Sco) and sensillum basiconicum 1 (Sb1).

The second segment (Lb2) is longer than the first segment (Table 1). Viewed from the ventral and dorsal sides, the base and ends are wider, while the middle part is narrower (Figure 4A,C). However, from the lateral view, the middle part is expanded, and both ends are narrowed (Figure 4B). Six types of sensilla were found on this segment, including three types of sensilla basiconica (Sb1, Sb2, Sb3), one type of sensilla campaniformia (Sca1), and one type of sensilla trichodea (St1) and sensilla coeloconica (Sco) (Figure 4D–H). Also, there are some cuticular pores (Cpo) arranged on the surface of second segment (Figure 4E).

The third segment (Lb3) is a little longer than the second, and of uniform width on both sides (Table 1, Figure 5A–C). Generally, there is a groove on the dorsal surface of the last 3/5 (Figure 5C). A wrinkled area is present on the dorsal surface at the internode between the second and third labial segment. Three types of sensilla are distributed on this part, including sensilla basiconica 1 (Sb1), sensilla trichodea (St1) and sensilla coeloconica (Sco) (Figure 5D–G).

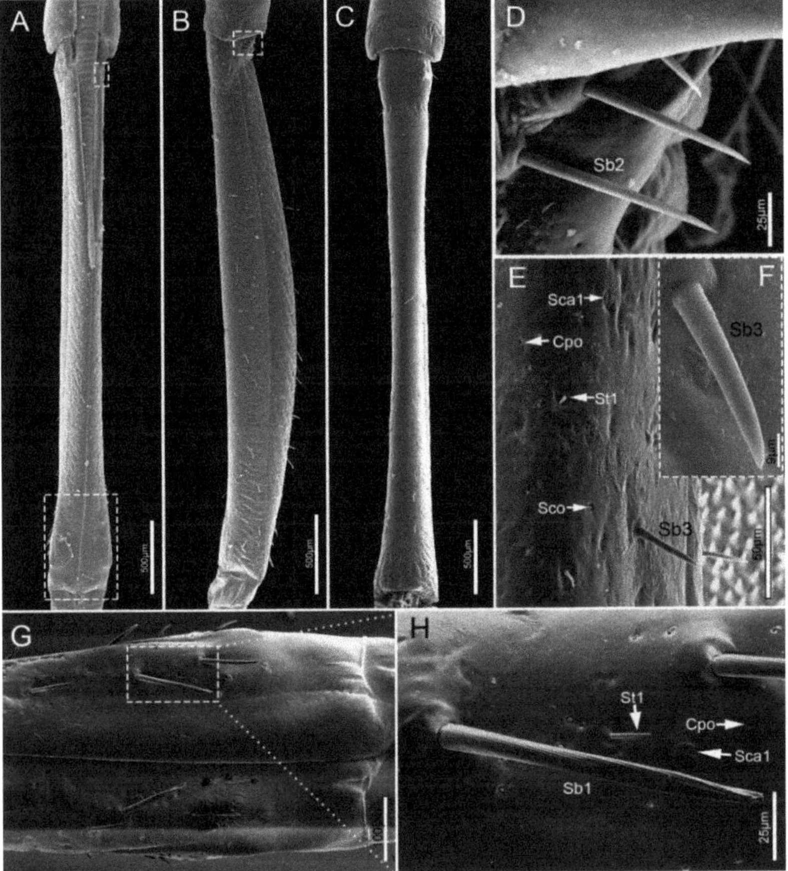

Figure 4. SEM of second labial segment of *E. fullo*. (**A**) Ventral view; (**B**) Lateral view; (**C**) Dorsal view; (**D**) Enlarged view of outlined box of (**B**) showing sensilla basiconica 2 (Sb2); (**E**) Enlarged view of outlined box of (**A**) showing sensilla campaniformia 1 (Sca1), cuticular pores (Cpo), sensilla trichodea 1 (St1), sensilla coeloconica (Sco) and sensilla basiconica 3 (Sb3); (**F**) Sensillum basiconicum 3 (Sb3); (**G**) Enlarged view of surface of the second segment of outlined box of (**A**); (**H**) Enlarged view of outlined box of (**G**), showing sensilla basiconica 1 (Sb1), sensilla trichodea 1 (St1), cuticular pore (Cpo) and sensilla campaniformia 1 (Sca1).

The fourth segment (Lb4) is nearly conical, of uniform width from base to apical 1/4 then narrowing to the apex (Figure 6A–C). There are abundant sensilla distributed on this segment, including three types of sensilla trichodea (St1, St2, St3), three types of sensilla basiconica (Sb1, Sb2, Sb3), two types of sensilla campaniformia (Sca1, Sca2) and sensilla coeloconica (Sco) (Figures 6D–I and 7A–E). *E. fullo* has very long and numerous sensilla trichodea 3 (St3) covering the end of the labium giving it a brush-like appearance (Figure 7A–C). Several sensilla basiconica 5 (Sb5) are visible among these sensilla trichodea 3 (St3) (Figure 8A,B).

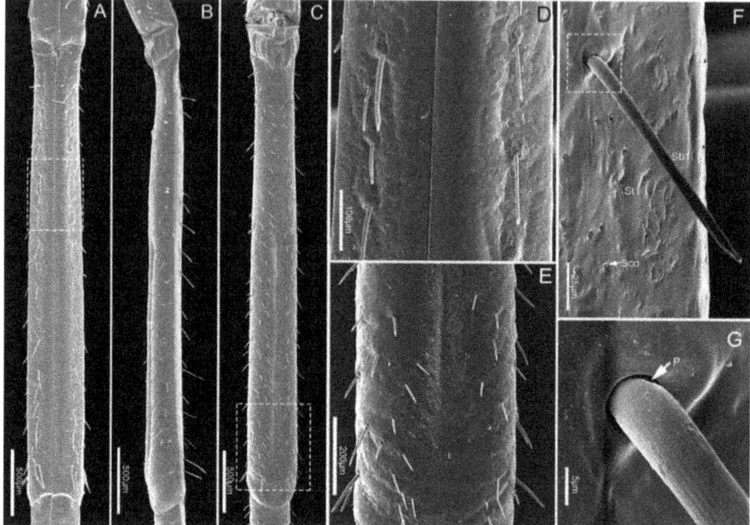

Figure 5. SEM of the third labial segment of *E. fullo*. (**A**) Ventral view; (**B**) Lateral view; (**C**) Dorsal view; (**D**) Enlarged view of outlined box of (**A**); (**E**) Enlarged view of outlined box of (**C**); (**F**) Enlarged view of surface of the third segment, showing sensilla basiconica 1 (Sb1), sensilla trichodea 1 (St1), and sensilla coeloconica (Sco); (**G**) Enlarged view of outlined box of (**F**), showing base pore (p).

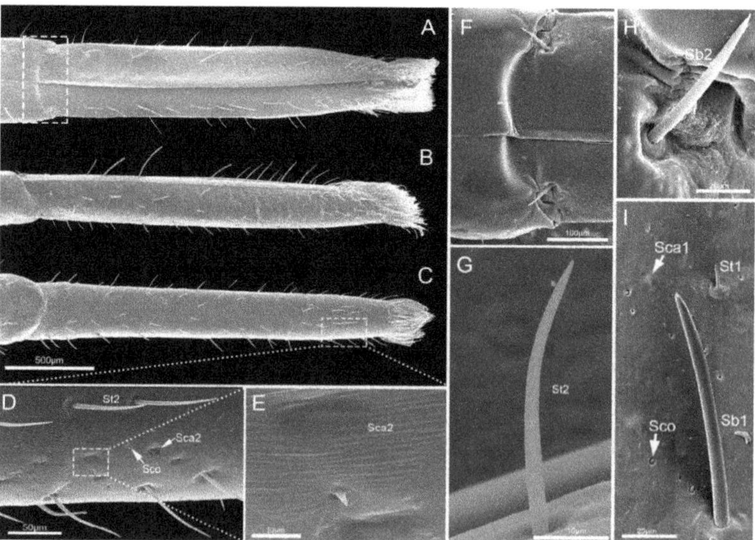

Figure 6. SEM of the fourth labial segment of *E. fullo*. (**A**) Ventral view; (**B**) Lateral view; (**C**) Dorsal view; (**D**) Enlarged view of outlined box of (**C**), showing sensilla trichodea 2 (St2), sensilla coeloconica (Sco) and sensilla campaniformia 2 (Sca2); (**E**) Enlarged view of outlined box of (**D**), showing sensilla campaniformia 2 (Sca2); (**F**) Enlarged view of outlined box of (**A**), showing sensilla basiconica 2 (Sb2); (**G**) Sensillum trichodeum 2 (St2); (**H**) Sensillum basiconicum 2 (Sb2); (**I**) Enlarged view of surface of the fourth segment, showing sensilla campaniformia 1 (Sca1), sensilla trichodea 1 (St1), sensilla coeloconica (Sco) and sensillum basiconicum 1 (Sb1).

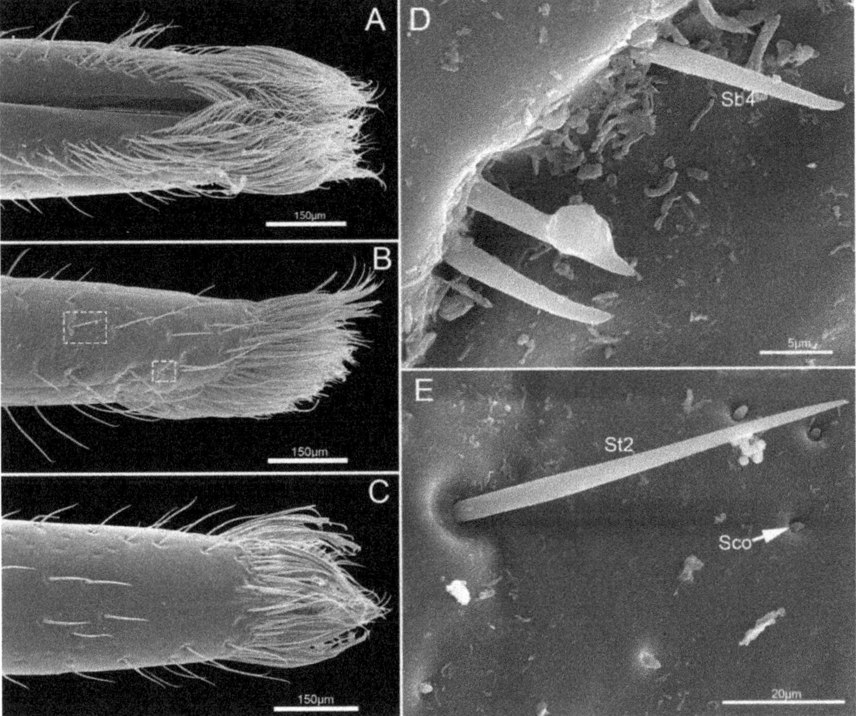

Figure 7. Proximal position the fourth labial segment of *E. fullo*. (**A**) Ventral view; (**B**) Lateral view; (**C**) Dorsal view; (**D**) Sensilla basiconica 4 (Sb4); (**E**) Enlarged view of outlined box of (**B**), showing sensillum trichodeum 2 (St2) and sensilla coeloconica (Sco).

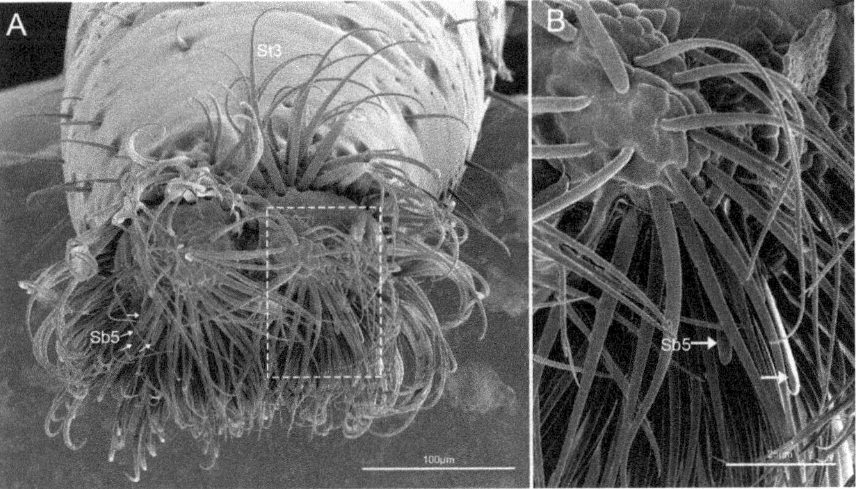

Figure 8. Tip of labium of *E.fullo*. (**A**) Vertical view of labial tip showing sensilla basiconica 5 (Sb5) and sensilla trichodea 3 (St3); (**B**) Enlarged view of outlined box of (**A**), showing sensilla basiconica 5 (Sb5).

3.1.3. Labial Sensilla Types and Their Arrangement

Based on their external morphology and distribution, eleven types (subtypes based on the length and shapes are distinguished) of distinct sensilla were observed on the surfaces of the labial segments. They were classified as: sensilla trichodea (St), sensilla campaniformia (Sca), sensilla coeloconica (Sco) and sensilla basiconica (Sb).

Sensilla trichodea (St) are hair-like sensilla. Their walls are smooth without any pores or grooves on the surface. Three subtypes of sensilla trichodea were distinguished. Sensilla trichodea 1 (St 1) are short (Table 2), aporous, smooth, with a slightly rounded tip and flexible sockets (Figure 2F). These sensilla are numerous and uniformly distributed on the labrum (Lm) and labium (Lb1–4) (Figures 2C and 3D). Sensilla trichodea 2 (St 2) are longer than sensilla trichodea 1 (Table 2), straight, with a smooth surface, a rounded tip and flexible sockets (Figure 6D,G). These sensilla are uniformly distributed on the fourth labial segment (Lb4). Sensilla trichodea 3 (St 3) are the longest sensilla (Table 2). These sensilla are curved at the tip and embedded in inflexible sockets. These sensilla are very numerous and located on the tip of the labium (Figure 8A,B).

Five subtypes of sensilla basiconica were distinguished. Sensilla basiconica 1 (Sb 1) are hair-like sensilla identical sensilla trichodea except for their smooth walls and blunt-tip. In the studied species, these sensilla are long (Table 2) ribbed and straight, slightly branched at the tip and arise from a cuticle with a flexible socket (Figure 2D,E). These sensilla are distributed on the labrum (Lm) and labium (Lb 2–4) (Figure 2B, Figure 3B,D,F, Figure 4G,H, Figure 5D–F and Figure 6I). Sensilla basiconica 2 (Sb 2) are cones with a smooth surface that arise from flexible sockets (Figures 4D and 6H). Three pairs of sensilla basiconica 2 are arranged at the junction of the first and second segment, two are present on each side of the junction of the third and fourth segments (Figures 4D and 6F). Sensilla basiconica 3 (Sb3) are short with a smooth surface, have a sharp tip and sit in a pit (Figure 4E,F). These sensilla are sparsely distributed on the ventral surface of the second segment. Sensilla basiconica 4 (Sb4) are peg-like with a smooth surface and have a rounded tip (Figure 7D). They are sparsely distributed on the lateral surface of the last segment (Figure 7B). Sensilla basiconica 5 (Sb 5) are present at the center of each distal lobe (Figure 7A). This type of sensillum is long, straight and with a smooth surface and a rounded tip, probably with a terminal pore (Figure 8B). Several of these sensilla basiconica are visible among the sensilla trichodea of the distal brush (Figure 7A,B).

Sensilla campaniformia (Sca) are flat, oval-shaped discs. Two subtypes of sensilla campaniformia are distinguished. Sensilla campaniformia 1 (Sca 1) are large (Table 2), numerous and present on the labrum (Lm) and labium (Lb 1–4) (Figure 2C, Figure 3E, Figure 4E,H and Figure 6I). Sensilla campaniformia 2 (Sca 2) are smaller than sensilla campaniformia 1 (Sca 1), fewer in number and located on the antero-lateral surface near the apical 1/3 (Figure 6D,E).

Sensilla coeloconica (Sco) consist of a small oval protuberance or cone inserted in a cuticular depression (Figure 2G). These are located on the labrum (Lm) and labium (Lb1–4) (Figure 2C, Figure 3C,D, Figure 4E, Figure 5F, Figure 6D,I and Figure 7E). These are without pores and have an inflexible socket.

Table 2. Distribution, morphometric data (mean ± SE), terminology and definition of sensilla used in the present paper. Data are mean ± SE values obtained from scanning electron microscopy. N = sample number; Lm, labrum; Lb, 1, 2, 3, 4, the first, second, third, fourth segment of labium; St 1–3, sensilla trichodea 1–3; Sb 1–5, sensilla basiconica 1–5; Sco, sensilla coeloconica; Sca 1–2, sensilla campaniformia 1–2; SF, sensory field on the labial tip; Wp, wall pore; Tp, tip pore.

Type	Location on Mouthparts	Length (μm)	Basal Diameter (μm)	N	Shape	Socket	Surface	Pore	Category	Function
St1	Lm, Lb1–4	12.0 ± 1.7	1.6 ± 0.2	10	Hair	Flexible	Smooth	No	Mechanoreceptive sensilla	Tactile
St2	Lb4	79.8 ± 1.9	4.92 ± 0.2	7	Hair	Flexible	Smooth	No	Mechanoreceptive sensilla	Tactile
St3	Lb4	83.1 ± 6.3	4.2 ± 0.6	16	Hair	Inflexible	Smooth	No	Mechanoreceptive sensilla	Tactile
Sb1	Lm, Lb1–4	89.0 ± 15.4	7.4 ± 1.2	20	Hair	Flexible	Grooved	No	Mechanoreceptive sensilla	Tactile
Sb2	Lb2, Lb 4	86.3 ± 7.1	6.5 ± 0.7	6	Peg	Flexible	Smooth	Wp (Uniporous)	Proprioceptive sensilla	Perceive the degree of flexion of the joint
Sb3	Lb2	55.8 ± 3.3	6.8 ± 0.4	16	Peg in pit	Inflexible	Smooth	No	Thermo-hygroreceptive sensilla	Temperature/humidity
Sb4	Lb4	12.8 ± 0.6	1.9 ± 0.1	6	Peg in pit	Inflexible	Smooth	No	Thermo-hygroreceptive sensilla	Temperature/humidity
Sb5	SF	Longer than Sb1	Wider than Sb1		Peg	Flexible	Smooth	Tp	Chemoreceptive sensilla	Gustatory
Sco	Lm, Lb1–4		2.4 ± 0.4	10	Pegs in cavity	Inflexible	Smooth	No	Thermo-hygroreceptive sensilla	Temperature/humidity
Sca1	Lb1, 2, 4		6.7 ± 0.4	4	Oval plate	Inflexible	Smooth	No	Proprioceptive sensilla	Perceive the degree of flexion of the joint
Sca2	Lb4		6.2 ± 1.2	4	Oval plate	Inflexible	Smooth	No	Proprioceptive sensilla	Perceive the degree of flexion of the joint

3.2. Stylet Fascicle

The stylet fascicle is long, slender, and composed of two separated mandibular stylets and two interlocked maxillary stylets (Figure 9A), ensheathed by the labium at rest and extending from the opening of the labial tip during feeding.

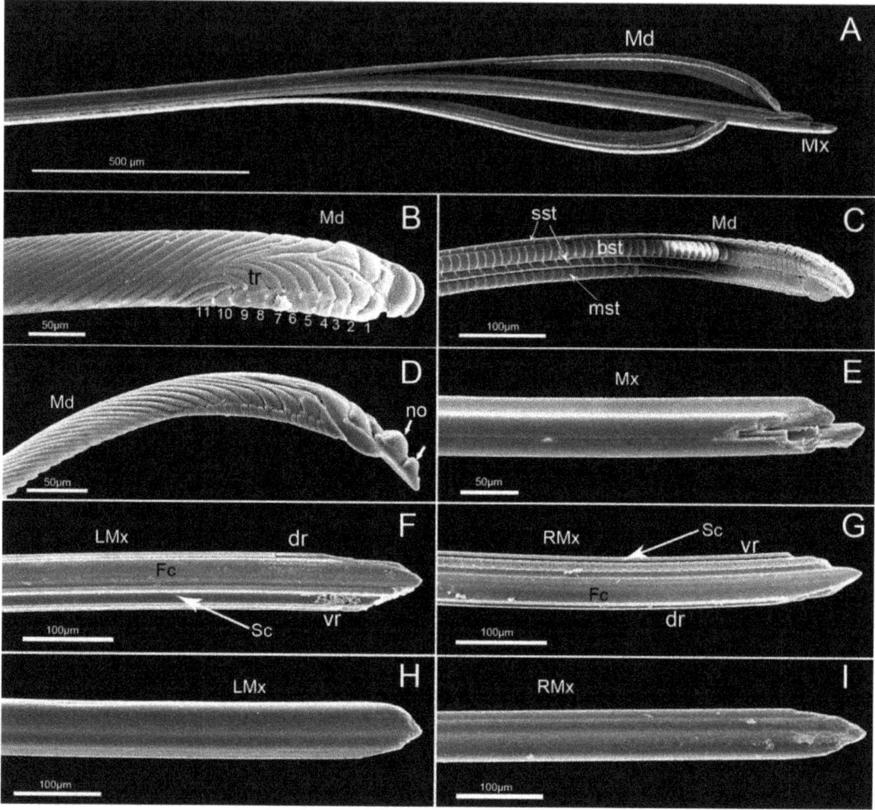

Figure 9. SEM of stylet fascicle of *E. fullo*. (**A**) Stylet fascicle showing mandibular (Md) and maxillary stylets (Mx); (**B**)External view of mandibular stylet (Md) showing eleven short transverse ridges (tr); (**C**) Interior view showing small squamous textures (sst), bigger squamous textures (bst) and middle squamous textures (mst); (**D**) Lateral view showing two nodules (no); (**E**) Apices of interlocked maxillary stylets; (**F**) Apex of left maxillary stylet (LMx) showing food canal (Fc) and salivary canal (Sc); (**G**) Apex of right maxillary stylet (RMx) showing food canal (Fc) and salivary canal (Sc); (**H**) External view of left maxillary stylet (LMx); (**I**) External view of right maxillary stylet (RMx); dr, dorsal side; vr, ventral side.

The mandibular stylets, located on each side of the maxillary stylets, are crescent-shaped in cross-section, convex externally and slightly concave internally to form a groove enclosing the maxillary stylets. On the lateral surface of each mandibular stylet, a series of approximately parallel, curved serrate ridges or teeth (a regular series of longer transverse ridges and eleven shorter transverse ridges) extend over the most distal part (Figure 9B). The most obvious features observed on the mandibular stylets of this species are two nodules present on the dorsal margin of the convex external surface near the apex (Figure 9B,D). There are four rows of squamous structures regularly distributed on the inner surface of the mandibular stylet (Figure 9C). The first and third rows consist of small squamous textures (sst), the second has bigger squamous textures (bst) and the fourth has medium-sized squamous textures (mst) with different cuticular spines.

The maxillary stylets (Mx) are interlocked by hook-like hinges and are not symmetrical (Figure 9E). The hook-like hinges include three joints from the cross-section, one of which is located at the center of the maxillary stylets and two of which are positioned at the lateral sides (Figure 10A,B). The external and internal surface of a maxillary stylet is smooth and the tip is sharp (Figure 9F–I). A row of nodes is present on the joint surface of the left stylet, which opposes the series of indentations on the right stylet (Figure 9G). A food canal (Fc) and salivary canal (Sc) are formed by the interlocked maxillary stylets, and the width of food canals is evenly distributed across the two stylets, while most of the salivary canal is housed in the right stylet (Figure 9F,G). The diameter of the central food canal is much greater than that of the salivary canal (Figure 10A,B). The cross-section of the stylet fascicle shows that each mandibular stylet has a dendritic canal, which is a large duct that runs the length of the stylet and is located centrally in the thickest portion of each structure (Figure 10A,B).

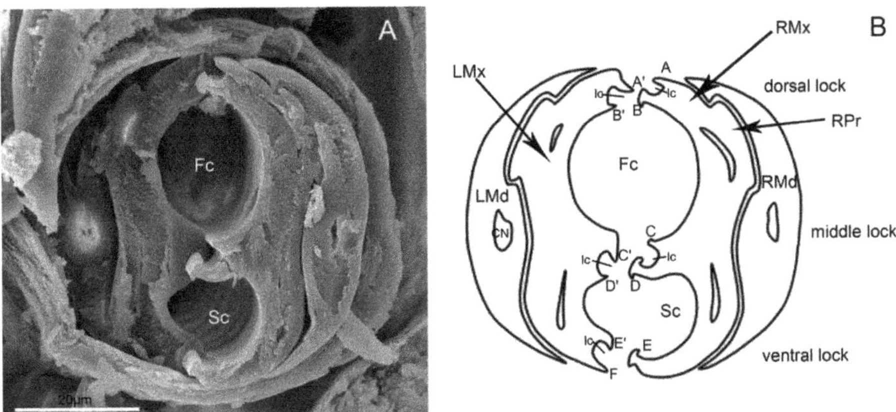

Figure 10. Cross-section of stylet fascicle of *E. fullo*. (**A**) Cross-section of stylet fascicle through middle of second and third segment showing food canal (Fc) and salivary canal (Sc); (**B**) Diagram of cross-section of stylet fascicle. LMd, left mandibular stylet; RMd, right mandibular stylet; LMx, left maxillary stylet; RMx, right maxillary stylet; Fc, food canal; Sc, salivary canal; Ic, interlocking canal; CN, dendritic canal; RPr, Right process of the maxilla; A, Straight; A′, Hooked; B, Hooked; B′, Straight; C, Straight; C′, Hooked; D, T-shaped; D′, Hooked; E, Hooked; E′, Hooked; F, Straight.

3.3. The Process of Feeding by E. fullo

The adult feeding process involves several steps, including the exploring and puncturing of the plant epidermis, a probing phase, an engorgement phase, and removal of the mouthparts from the plant tissue. These processes vary slightly in mouthpart position and duration.

When the insect is at rest or not feeding, the rostrum is in contact with the ventral surface of the body from the front coxal base to the anterior part of the abdomen (Figure 11A). The proximal end of labial segment 2 articulates with the bucculae.

Insects feeding on the internal fluids of other organisms must first penetrate the plant tissues. After landing, an adult of *E. fullo* walks on its plant and explores for a suitable feeding location by probing. It then stops and remains still while the antennae swing up and down several times. After a few seconds, gripping the plant with its legs, the bug tilts the anterior part of its body upward at an angle to the surface, and the rostrum is then extended forward and used as a sense organ in conjunction with the eyes and antennae to examine the plant material for a suitable feeding spots. The labium first moves forward by swinging from its horizontal position of repose until it is perpendicular to the plant surface. The rostrum tip then taps the surface or slides over it. When the labium is moved forward from its resting position, the stylet tip reaches the tip of the labium and may even extend a short distance beyond the top (Figure 11B).

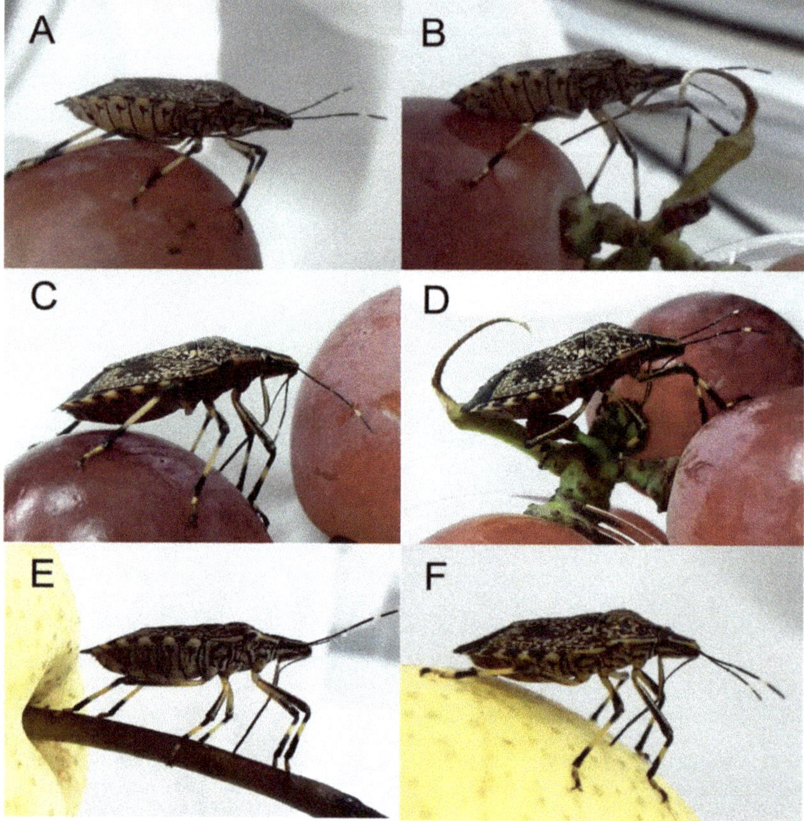

Figure 11. Feeding on fruits and young stalks in adult *E. fullo* showing positions of the mouthparts. (**A**) At rest or not feeding; (**B**) Exploring suitable feeding location; (**C**) Feeding on grape; (**D**) Feeding on green stalk of grape; (**E**) Feeding on stalk of pear; (**F**) Feeding on pear.

Upon contact with a potential feeding site, the bug may probe with sensilla on the tip of the labium and penetrate the site with the stylets to test if this site is suitable for feeding. After selecting an appropriate feeding site, the insect then presses the tip of the labium onto the plant surface and inserts the feeding stylets. Then, the labium makes an elbow-like bend between the first and second segment, while the base of the stylet fascicle is held in the groove of the labrum (Figure 12A). The labium continues retracting to its maximum extent, at which the angle between the first and second segments is nearly 90°, allowing the head to be lowered as the stylet bundle penetrates the food tissue (Figure 12B,C), with the maxillary stylets lagging slightly behind the mandibular stylets. Stylet probing continues until a suitable tissue is found. It may take anywhere from five minutes to three hours from the beginning of probing until a feeding site is reached. The bug secretes viscous saliva as the stylets progress through the tissue.

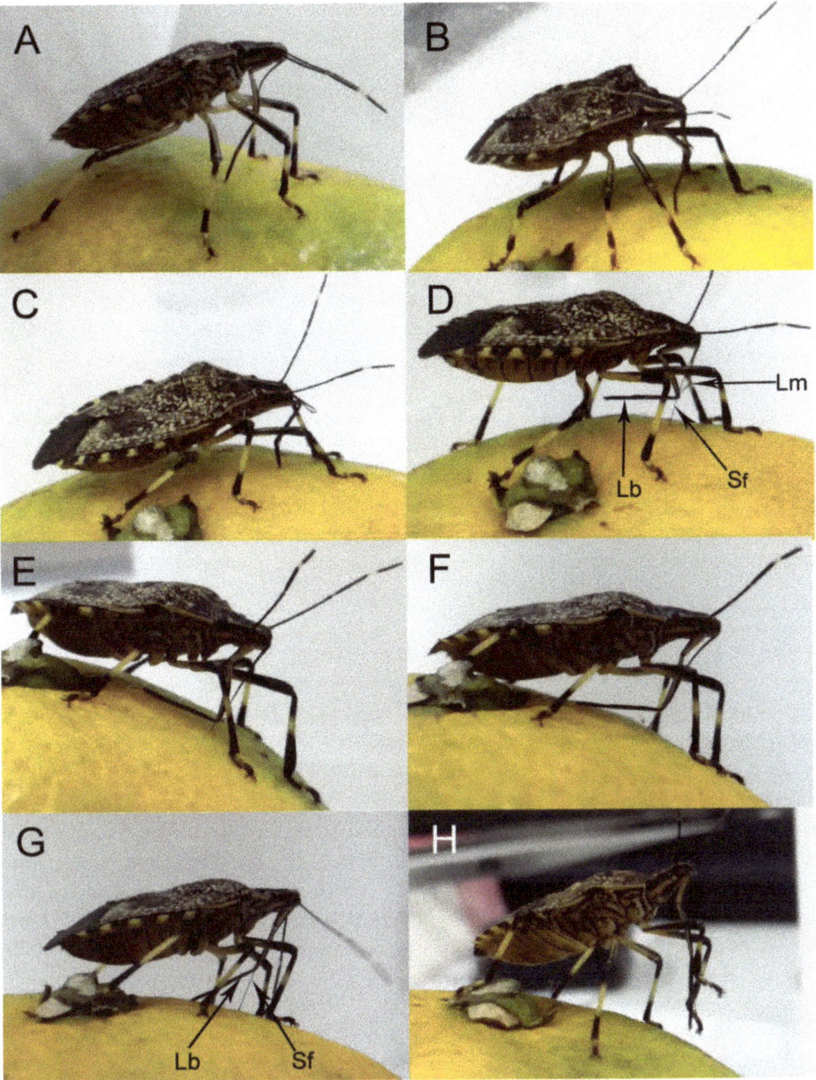

Figure 12. Feeding stages on orange in adult *E. fullo* showing positions of the mouthparts. (**A**) Location of suitable feeding position by the labium; (**B**), (**C**) Puncture of orange by stylet fascicle showing elbow-like fold of proximal and second rostral segments and stylet penetration; (**D**) The bug lifts up the third and fourth segments of labium parallel to host surface and then feeds; (**E**) The bug gradual straightens the first and second labial segments; (**F**), (**G**) Termination of feeding showing retraction of stylets; (**H**) Use of forelegs to return stylet fascicle to labial groove; Lm, labrum; Lb, labium; Sf, stylet fascicle.

After a feeding site is reached, the bug bends the third and fourth segments of the labium backward, away from the inserted stylets until the distal section of the labium is parallel to the host surface, after which feeding can commence (Figure 12D). The bug then extracts and sucks host fluids repeatedly. Feeding may last from a few seconds to one hour at a time.

When finished feeding, the bug gradually straightens the first and second labial segments; meanwhile, the third and fourth labial segments rotate forward and contact the host surface (Figure 12E). Then the body gradually raises and pulls out the stylet fascicle (Figure 12F,G). The bug replaces the stylet fascicle into the labial groove with the help of the forelegs (Figure 12H). Finally, the rostrum rotates back to its resting position along the sternum.

The process of feeding on young shoots of the plant is similar to that observed for fruit feeding, except that the stylets are never fully retracted from the labium (Figure 11C–F).

4. Discussion

Substantial data are available on structure and function of mouthparts in Hemiptera. However, detail on the mechanics of feeding behavior, especially with respect to the sensory and motor feedback mechanisms, is lacking [45–47]. A study of the fine morphology of mouthparts allows us to interpret the function of the component parts of the feeding apparatus and improves our understanding of the actual feeding mechanism.

In this study, the feeding behavior of *E. fullo* is described. To our knowledge, this is the first time that the detailed mouthpart morphology and feeding performance in a member of Pentatomidae have been reported together. The modified mouthparts of *E. fullo* have a number of morphological similarities to those heteropteran species described previously [17,19,22,37,44,48–55], but our study revealed some new and interesting features that differ from those of other true bugs, and provide a better understanding of the feeding strategies and the sensory systems of *E. fullo*.

4.1. Mouthpart Morphology and Their Adaptability to Feeding

The labrum, a conspicuous anterior structure on the adult insect head, should play an important role in insect feeding. Recently the labrum was reinterpreted as fused paired appendages of an intercalary segment [56–58] and a few scholars have conducted detailed studies on its morphology in Heteroptera [52–55,59–61]. In previously published reports, the morphology of the labrum was used as a taxonomic feature of higher taxa of Heteroptera [59–63], but its structure varies according to feeding habits and mechanisms [61]. Spooner [59] recognized three basic types of labrum in Heteroptera: (1) a broad, flap-like labrum; (2) a long, narrow, triangular labrum; (3) a broad, flap-like sclerite with a long epipharyngeal projection. The labrum of *E. fullo* corresponds to the second group. This is similar to other true bugs, e.g., *Pyrrhocoris sibiricus* [52], *Cheilocapsus nigrescens* [53], and four species of Largidae (*Physopelta quadriguttata*, *Ph. gutta*, *Ph. cincticallis*, and *Macrocheraia grandis*) [55]. We observed regular wrinkles from base to end on the ventral surface of the *E. fullo* labrum. These wrinkles may function to add flexibility to the labrum, allowing deeper stylet penetration (Figure 12B,C). Such a long labrum (Table 1) may also be used to hold the basal part of the stylet fascicle in the labial groove during feeding (Figure 12D).

The labium of *E. fullo* has four segments as in most of other heteropteran bugs [27,52–55]. Usually, when insects are feeding, the second segment approaches the first segment, allowing the head to be lowered as the stylet fascicle penetrates the food tissue. Previous studies have reported that a band-like dorsal plate is present between the first and the second segment [27,52,54,55,64–66], while the base of the stylet fascicle is held in the groove of the labrum. However, in our study, there was no such structure (a band-like dorsal plate), and the distal part of the dorsum of the first segment is contracted inward (Figures 11 and 12). This is probably because, unlike *Pyrrhocoris sibiricus* [52], which moves the labium back to its abdomen, *E. fullo* bends the first and second segments for deeper feeding. Moreover, the first and second segment are stronger than the third and fourth segments. The first and second labial segments of *E. fullo* are presumably folded to support the head, allowing the stylet fascicle to penetrate the plant (Figure 12B).

Heteropteran stylets form a fascicle composed of two lateral mandibular stylets and two maxillary stylets; the former are armed with teeth or rasps and the latter interlock and forms the salivary and food canals [17,18,25]. As feeding and probing on host plants are responsible for the direct or indirect damage to plants by phytophagous hemipteran insects, the stylets, including the shape and dentition of the tips, have been studied previously in several heteropterans [17,18,20,49,50,52–55,67–75]. In *E. fullo*, there are a series of squamous textures regularly distributed on the inner surface of the mandibular stylet and the left and right sides of the longitudinal groove are different. Similar structures are found in other phytophagous species [17,52,53,55]. Cobben [17] mentioned that the orientation of this parallel groove is such that the forward thrust of one mandible will cause considerable friction against the outer surface of the adjacent maxillary stylet contributing to its inward deviation. We also observed two nodules present on the dorsal margin of the external surface and a series of transverse ridges arranged on the outer surface. In different phytophagous Heteroptera, the nodules are slightly different, and the number of nodules also varies [49,52–55]. Depieri and Panizzai [49] observed 1 to 4 central teeth and 1–3 lateral teeth in *Dichelops melacanthus, Euschistus heros, Nezara viridula* and *Piezodorus guildinii*. Wang and Dai [52] found that mandibular stylets of *P. sibiricus* have three central teeth and two paired lateral teeth on the distal extremity, as well as five or six oblique parallel ridges on the subapex of the external convex region. In polyphagous species of Largidae (*Physopelta quadriguttata, Ph. gutta, Ph. cincticallis,* and *Macrocheraia grandis*), the serration pattern of the mandibles is 1–3 central teeth and 1–2 lateral teeth [55]. The teeth at the tip of the mandibular stylet may help to fix the stylets in host tissues [17,76].

Both mandibles together with the maxillary bundle function as a single plunging instrument [17]. Maxillary stylets are asymmetrical only in the internal positions of the longitudinal carinae and grooves. Their inner surfaces show traces of small, widely spaced notches arranged in longitudinal strips. As found by Cobben [17] in his study of *Graphosoma lineatum* L., we also found these grooves on the maxillary stylets of *E. fullo* to form a salivary canal (Sc) and a food canal (Fc). The maxillary stylets are longer than the mandibular stylets and the salivary canal is narrower than the food canal as in *Pyrrhocoris sibiricus* [52], *Cheilocapsus nigrescens* [53], *Stephanitis nashi* [54] and four species of Largidae (*Physopelta quadriguttata, Ph. gutta, Ph. cincticallis,* and *Macrocheraia grandis*) [55]. In *E. fullo*, the maxillary stylets are smooth externally but equipped with a longitudinal ridge that engages grooves in the mandibular stylets, causing it to curve inward during probing of plant tissue [17]. Moreover, the sharp ends of the maxillary stylet are specialized to pierce plant tissues while probing.

Brożek and Herczek [37] have studied the interlocking mechanisms of maxillae and mandibles in Heteroptera. Three locks between maxillae and mandibulae have been identified, i.e., dorsal, middle and ventral, similar to Fulgoroidea [3], in contrast with two locks in leafhoppers [6,8]. Our observation of the internal structure of *E. fullo* mouthparts based on the cross-section of the subapical segment of the rostrum reveals the same number of processes in each of the three locks. The food canal is oval and the salivary canal is smaller than that of the food canal, which is semicircular in cross-section. Both maxillary and mandibular stylets are flattened laterally; thus they are higher than wide in cross-section [37]. There are five upper processes on the right maxilla and six processes on the left maxilla, as found by Brożek and Herczek [37] in their study of other representatives of the Pentatomidae, e.g., *Acanthosoma haemorrhoidale* and *Elasmucha fieberi*.

Heteropteran insects have four feeding methods including stylet-sheath feeding, lacerate-and-flush feeding, macerate-and-flush feeding and osmotic pump feeding [17,28,31], and each is used on a different kind of host tissue. Miles [77] suggested that some pentatomomorphans can employ two types of feeding and that both phytophagous or carnivorous Pentatomorpha produce a stylet sheath. Generally, the polyphagous *E. fullo* primarily suck sap from the trunk, leaves, immature stems and fruits. Therefore, this species presumably employs the stylet-sheath feeding method when feeding from the phloem of the host plant, and employs lacerate-and-flush feeding when feeding on the fruit. In the stylet-sheath feeding method, the insect inserts the stylets into the feeding site (mainly phloem) and forms a salivary sheath around the stylets. In the lacerate-and-flush feeding type, these insects use their strong mandibular teeth to lacerate cells and the sharp ends of the maxillary stylet to pierce fruit for flush and suck feeding.

Usually, before feeding, heteropteran insects secrete some saliva on the surface of the host plant which is re-absorbed repeatedly to test the suitability of the feeding site [78]. In our observations, the labium lip of *E. fullo* has abundant sensilla trichodea (St3) and few sensilla basiconica (Sb5). We suspect that these large numbers of sensilla trichodea (St3) may be used to smear the saliva and the sensilla basiconica (Sb5) act as chemical sensors to taste the liquid.

4.2. Labial Sensillar System

Many previous authors have described rostral sensilla of Hemiptera and their possible function as chemoreceptors and mechanoreceptors [4,22,48,52–55,79,80]. Detailed morphological descriptions of Pentatomidae labial sensilla have never been previously reported. In this study, eleven types of sensilla were observed on the mouthparts of *E. fullo*.

The sensilla that cover the labial surface (except the labial tip) in *E. fullo* are evidently similar to those of most pentatomomorphan species, as well to other heteropteran species, as reported by several authors [22,44,52–55]. According to the inferred functions of the sensilla, we divided the sensilla on the labial surface into three categories: Thermo-hygroreceptive, proprioceptive and mechanosensory [43,81]. Mechanosensory sensilla include sensilla trichodea (St1, St2) and sensilla basiconica (Sb1), which have no pores or are uniporous and are embedded in flexible sockets. The proprioceptive sensilla include sensilla basiconica (Sb2), located on the junction between the first and second labial segment, and the third and fourth segment, and nonporous cupola (Sca1, Sca2) located on the surface of the cuticle or enclosed in a pit. The thermo-hygroreceptive sensilla include five types (Sb3, Sb4, Sco). Generally, all of the sensilla with this function are nonporous pegs (Sb3, Sb4, Sco).

The labial tip, which contacts the host surface during host selection and feeding, usually has poreless mechanosensory hairs and uniporous or multiporous pegs [34]. According to Rani [26], the carnivorous stinkbug *Eocanthecona furcellata* (Wolff) possesses numerous sensilla of different types at the tip of the labium, e.g., trichoid sensilla, long hairs with profusely branched shafts, an oval-shaped peg surrounded by sensory hairs with branched shafts and a short, stout peg encircled by a group of long hair-like sensilla. Six types of labial sensilla on the labium of phytophagous and predatory pentatomid species were described by Shama et al. [44]. Both studies found long cuticular projections and no sensory function on the labial tip. Nevertheless, in this study we observed in *E. fullo* many very long sensilla trichodea (St3) covering the labial tip, as well as a few sensilla basiconica (Sb5) on the central tip of the labium. Sensilla trichodea probably represent mechanosensilla as their morphology suggests, whereas sensilla basiconica are gustatory (chemosensitive sensilla). *E. fullo* is a polyphagous species sucking the sap from leaves, immature stems and fruits similar to other pentatomomorphan species. Feeding by this species causes yellowish brown spots to appear on the surface of the plant. Extensive injury results the leaf falling off. Damage to fruts includes which causes loss of edible value and yield loss [40]. So far, this is the only pentatomid species observed to have such long and numerous sensilla of the labial tip. Other studied polyphagous heteropteran species have fewer such sensilla (10 to 12 sensilla) and are more uniform in structure [22,44,52–55,82,83] in contrast to *E. fullo* in which the sensilla are much more numerous.

5. Conclusions

To sum up, the feeding structures in the few species of Pentatomidae studied so far seem similar to each other, presumably due to strong structural and functional constraints on their evolution. However, the mouthparts of *E. fullo* differ from those of previously studied stink bugs in the cross-sectional shape of the stylets, arrangement of labial sensilla and number of teeth of the mandibular stylets. This dissimilarity from other species of Pentatomidae and species of other hemipteran families so far described makes *Erthesina fullo* unique, particularly in its excessively long and numerous sensilla trichodea covering the end of labium. The structure and function of the mouthparts of this species are adapted for phytophagous feeding habits.

The adult feeding process involves several steps, including the exploring and puncturing of the host epidermis, a probing phase, an engorgement phase, and removal of the mouthparts from the host tissue. Studies of feeding behavior and mouthpart morphology of additional pentatomid species are needed to determine how much variation occurs in this diverse and economically important family.

Author Contributions: Data curation, W.D.; funding acquisition, W.D.; investigation, Y.W. and W.D.; project administration, W.D.; resources, W.D.; writing – original draft, Y.W. and W.D. All authors have read and agreed to the published version of the manuscript.

Funding: This project was supported by the National Natural Science Foundation of China (Nos. 31772514, 31572306, 31272343) and the Program of the Ministry of Science and Technology of the People's Republic of China (2015FY210300).

Acknowledgments: We thank John Ri.chard Schrock (Emporia State University, Emporia, KS, USA) and Chris Dietrich (Illinois Natural History Survey, Champaign, IL, USA) for his comments on an earlier draft of this paper. We thank the Life Science Research Core Services of Northwest A&F University for providing scanning electron microscope.

Conflicts of Interest: The authors declare no conflict of interest.

References

1. Marais, E.; Klok, C.J.; Terblanche, J.S.; Chown, S.L. Insect gas exchange patterns: A phylogenetic perspective. *J. Exp. Biol.* **2005**, *208*, 4495–4507. [CrossRef] [PubMed]

2. Weintraub, P.G.; Beanland, L. Insect vectors of phytoplasmas. *Annu. Rev. Èntomol.* **2006**, *51*, 91–111. [CrossRef] [PubMed]

3. Brożek, J.; Bourgoin, T.; Szwedo, J. The interlocking mechanism of maxillae and mandibles in Fulgoroidea (Insecta: Hemiptera: Fulgoromorpha). *Polskie Pismo Entomologiczne* **2006**, *75*, 239–253.

4. Brożek, J.; Bourgoin, T. Morphology and distribution of the external labial sensilla in Fulgoromorpha (Insecta: Hemiptera). *Zoomorphology* **2012**, *132*, 33–65. [CrossRef]

5. Tavella, L.; Arzone, A. Comparative morphology of mouth parts of *Zyginidia pullula*, *Empoasca vitis*, and *Graphocephala fennahi* (Homoptera, Auchenorrhyncha). *Ital. J. Zool.* **1993**, *60*, 33–39. [CrossRef]

6. Leopold, R.A.; Freeman, T.P.; Buckner, J.S.; Nelson, D.R. Mouthpart morphology and stylet penetration of host plants by the glassy-winged sharpshooter, *Homalodisca coagulata*, (Homoptera: Cicadellidae). *Arthropod Struct. Dev.* **2003**, *32*, 189–199. [CrossRef]

7. Wiesenborn, W.D. Mouth parts and alimentary canal of *opsius stactogalus* Fieber (Homoptera: Cicadellidae). *J. Kans. Èntomol. Soc.* **2004**, *77*, 152–155. [CrossRef]

8. Zhao, L.; Dai, W.; Zhang, C.; Zhang, Y. Morphological characterization of the mouthparts of the vector leafhopper *Psammotettix striatus* (L.) (Hemiptera: Cicadellidae). *Micron* **2010**, *41*, 754–759. [CrossRef]

9. Ammar, E.-D.; Hall, D. New and simple methods for studying hemipteran stylets, bacteriomes, and salivary sheaths in host plants. *Ann. Èntomol. Soc. Am.* **2012**, *105*, 731–739. [CrossRef]

10. Pollard, D.G. Some aspects of plant penetration by *Myzits persicae* (Sulz.) nymphs (Homoptera, Aphididae). *Bull. Entomol. Res.* **1971**, *61*, 315–324. [CrossRef]

11. Pollard, D.G. Plant penetration by feeding aphids (Hemiptera, Aphidoidea): A review. *Bull. Èntomol. Res.* **1973**, *62*, 631–714. [CrossRef]

12. Tjallingii, W.F. Mechanoreceptors of the aphid labium. *Èntomol. Exp. Appl.* **1978**, *24*, 731–737. [CrossRef]

13. Uzest, M.; Gargani, D.; Dombrovsky, A.; Cazevieille, C.; Cot, D.; Blanc, S. The "acrostyle": A newly described anatomical structure in aphid stylets. *Arthropod Struct. Dev.* **2010**, *39*, 221–229. [CrossRef] [PubMed]

14. Ahmad, A.; Kaushik, S.; Ramamurthy, V.; Lakhanpaul, S.; Ramani, R.; Sharma, K.; Vidyarthi, A. Mouthparts and stylet penetration of the lac insect *Kerria lacca* (Kerr) (Hemiptera:Tachardiidae). *Arthropod Struct. Dev.* **2012**, *41*, 435–441. [CrossRef] [PubMed]

15. Walker, G.P.; Gordh, G. The occurrence of apical labial sensilla in the Aleyrodidae and evidence for a contact chemosensory function. *Èntomol. Exp. Appl.* **1989**, *51*, 215–224. [CrossRef]

16. Rosell, R.C.; Lichty, J.E.; Brown, J.K. Ultrastructure of the mouthparts of adult sweet potato whitefly, *Bemisia tabaci* Gennadius (Homoptera: Aleyrodidae). *Int. J. Insect Morphol. Embryol.* **1995**, *24*, 297–306. [CrossRef]

17. Cobben, R.H. Evolutionary trends in Heteroptera. In *Part II Mouthpart-Structures and Feeding Strategies*; Mededelingen Landbouwhogeschool: Wageningen, The Netherlands, 1978; Volume 78, pp. 1–407.

18. Boyd, D.W. Digestive enzymes and stylet morphology of *Deraeocoris nigritulus* (Uhler) (Hemiptera: Miridae) reflect adaptations for predatory habits. *Ann. Èntomol. Soc. Am.* **2003**, *96*, 667–671. [CrossRef]

19. Anderson, W.G.; Hengmoss, T.M.; Baxendale, F.P.; Baird, L.M.; Sarath, G.; Higley, L. Chinch bug (Hemiptera: Blissidae) mouthpart morphology, probing frequencies, and locations on resistant and susceptible germplasm. *J. Econ. Entomol.* **2006**, *99*, 212–221. [CrossRef]

20. Kumar, S.M.; Sahayaraj, K. Gross morphology and histology of head and salivary apparatus of the predatory bug, *Rhynocoris marginatus*. *J. Insect Sci.* **2012**, *12*, 1–12. [CrossRef]

21. Garzo, E.; Bonani, J.; Lopes, J.R.S.; Fereres, A. Morphological description of the mouthparts of the Asian citrus psyllid, *Diaphorina citri* Kuwayama (Hemiptera: Psyllidae). *Arthropod Struct. Dev.* **2012**, *41*, 79–86. [CrossRef]

22. Rani, P.U.; Madhavendra, S.S. Morphology and distribution of antennal sense organs and diversity of mouthpart structures in *Odontopus nigricornis* (Stål) and *Nezara viridula* L. (Hemiptera). *Int. J. Insect Morphol. Embryol.* **1995**, *24*, 119–132. [CrossRef]

23. Freeman, T.P.; Buckner, J.S.; Nelson, D.R. Stylet length of whitefly adults and nymphs and the mechanism of stylet insertion into the leaves of host plants. *Microsc. Microanal.* **2000**, *6* (Suppl. 2), 876–877. [CrossRef]

24. Freeman, T.P.; Buckner, J.S.; Nelson, D.R.; Chu, C.C.; Henneberry, T.J. Stylet penetration by *Bemisia argentifolii* (Homoptera: Aleyrodidae) into host leaf tissue. *Ann. Entomol. Soc. Am.* **2001**, *94*, 761–768. [CrossRef]

25. Boyd, D.W.; Cohen, A.C.; Alverson, D.R. Digestive enzymes and stylet morphology of *Deraeocoris nebulosus* (Hemiptera: Miridae), a predacious plant bug. *Ann. Èntomol. Soc. Am.* **2002**, *95*, 395–401. [CrossRef]

26. Rani, P.U. Sensillary morphology on the rostral apex and their possible role in prey location behaviour of the carnivorous stinkbug, *Eocanthecona furcellata* (Wolff) (Heteroptera: Pentatomidae). *Acta Zool.* **2009**, *90*, 246–253. [CrossRef]

27. Esquivel, J.F. Estimating potential stylet penetration of southern green stink bug—A mathematical modeling approach. *Èntomol. Exp. Appl.* **2011**, *140*, 163–170. [CrossRef]

28. Backus, E.A. Sensory systems and behaviours which mediate hemipteran plant-feeding: A taxonomic overview. *J. Insect Physiol.* **1988**, *34*, 151–165. [CrossRef]

29. Strong, F.E. Physiology of Injury Caused by *Lygus hesperus*. *J. Econ. Entomol.* **1970**, *63*, 808–814. [CrossRef]

30. Tingey, W.M.; Pillimer, E.A. Lygus Bugs: Crop Resistance and Physiological Nature of Feeding Injury. *Bull. Entomol. Soc. Amer.* **1977**, *23*, 277–287. [CrossRef]

31. Hori, K. Possible causes of disease symptoms resulting from the feeding of phytophagous heteroptera, Chapter, 2. In *Heteroptera of Economic Importance*; Schaefer, C.W., Panizzi, A.R., Eds.; CRC Press: Boca Raton, FL, USA, 2000; pp. 11–35.

32. Mitchell, P.L. Heteroptera as vectors of plant pathogens. *Neotropical Èntomol.* **2004**, *33*, 519–545. [CrossRef]

33. Panizzi, A.R.; McPherson, J.E.; James, D.G.; Javahery, M.; McPherson, R.M. Stink bugs (Pentatomidae), Chapter, 13. In *Heteroptera of Economic Importance*; Schaefer, C.W., Panizzi, A.R., Eds.; CRC Press: Boca Raton, FL, USA, 2000; pp. 421–474.

34. Brożek, J. Labial sensillae and the internal structure of the mouthparts of *Xenophyes cascus* (Bergroth 1924) (Peloridiidae: Coleorrhyncha: Hemiptera) and their significance in evolutionary studies on the Hemiptera. *Aphids Hemipterous Insects* **2007**, *13*, 35–42.

35. Brożek, J. Morphology and arrangement of the labial sensilla of the water bugs. *Bull. Insectol.* **2008**, *61*, 67–168.

36. Baker, G.T.; Chen, X.; Ma, P.W. Labial tip sensilla of *Blissus leucopterus* (Hemiptera: Blissidae): Ultrastructure and behavior. *Insect Sci.* **2008**, *15*, 271–275. [CrossRef]

37. Brożek, J.; Herczek, A. Internal structure of the mouthparts of true bugs (Hemiptera: Heteroptera). *Pol. J. Entomol.* **2004**, *73*, 79–106.

38. Li, Q.C.; Cheng, A.Y.; Wang, H.S.; Zhang, W.Y. Control technicians of the *Halyomorpha picus* (Fabricius) and *Erthesina fullo* (Thunberg). *Plant Doctor* **1998**, *11*, 17–18. (In Chinese)

39. Mi, Q.Q.; Zhang, J.P.; Gould, E.; Chen, J.H.; Sun, Z.T.; Zhang, F. Biology, Ecology, and Management of *Erthesina fullo* (Hemiptera: Pentatomidae): A Review. *Insects* **2020**, *11*, 346. [CrossRef] [PubMed]

40. Song, H.W.; Wang, C.M. Studies on the harm and prevention of the *Halyomorpha picus* (Fabricius) and *Erthesina fullo* (Thunberg) to Chinese jujube. *Chin. J. Appl. Entomol.* **1993**, *30*, 225–228. (In Chinese)

41. Sun, S.X. Studies on the alimentary canal of *Erthesina fullo* Thunberg. *J. Shandong Agric. Coll.* **1956**, *2*, 37–50. (In Chinese)

42. Zhang, C.T.; Li, D.L.; Su, H.F.; Xu, G.L. A study on the biological characteristics of *Halyomorpha picus* and *Erthesina fullo*. *For. Res.* **1993**, *6*, 271–275. (In Chinese)

43. Altner, H.; Prillinger, L. Ultrastructure of invertebrate chemo, thermo, and hygroreceptors and its functional significance. *Int. Rev. Cytol.* **1980**, *67*, 69–139. [CrossRef]

44. Parveen, S.; Ahmad, A.; Brożek, J.; Ramamurthy, V.V. Morphological diversity of the labial sensilla of phytophagous and predatory Pentatomidae (Hemiptera: Heteroptera), with reference to their possible functions. *Zootaxa* **2015**, *4039*, 359–372. [CrossRef] [PubMed]

45. Backus, E.A. Anatomical and sensory mechanisms of leafhopper and planthopper feeding behavior. In *The Leafhoppers and Planthoppers*; Nault, L.R., Rodriguez, J.G., Eds.; John Wiley & Sons: New York, NY, USA, 1985; pp. 163–194.

46. Feir, D.; Beck, S.D. Feeding behavior of the large milkweed bug, *Oncopellus fasciatus*. *Ann. Èntomol. Soc. Am.* **1963**, *56*, 224–229. [CrossRef]

47. Saxena, K.N. Mode of ingestion in a heteropteran insect *Dysdercus koenigii* (F.) (Pyrrhocoridae). *J. Insect Physiol.* **1963**, *9*, 47–71. [CrossRef]

48. Rani, P.U.; Madhavendra, S.S. External morphology of antennal and rostral sensillae in four hemipteran insects and their possible role in host plant selection. *Int. J. Trop. Insect Sci.* **2005**, *25*, 198–207. [CrossRef]

49. Depieri, R.A.; Panizzi, A.R. Rostrum length, mandible serration, and food and salivary canals areas of selected species of stink bugs (Heteroptera, Pentatomidae). *Revista Brasileira De Entomologia* **2010**, *54*, 584–587. [CrossRef]

50. Depieri, R.A.; Siqueira, F.; Panizzi, A.R. Aging and food source effects on mandibular stylets teeth wear of Phytophagous Stink Bug (Heteroptera: Pentatomidae). *Neotrop. Èntomol.* **2011**, *39*, 952–956. [CrossRef]

51. Barsagade, D.; Gathalkar, G. First predation record of *Canthecona furcellata*(Wolff.) (Hemiptera: Pentatomidae) on spinning stage silkworm *Antheraea mylitta*(Drury). *Èntomol. Res.* **2016**, *46*, 236–245. [CrossRef]

52. Wang, Y.; Dai, W. Fine structure of mouthparts and feeding performance of *Pyrrhocoris sibiricus* Kuschakevich with remarks on the specialization of sensilla and stylets for seed feeding. *PLoS ONE* **2017**, *12*, e0177209. [CrossRef]

53. Wang, Y.; Li, L.; Dai, W. Fine Morphology of the mouthparts in *Cheilocapsus nigrescens* (Hemiptera: Heteroptera: Miridae) reflects adaptation for phytophagous habits. *Insects* **2019**, *10*, 143. [CrossRef]

54. Wang, Y.; Brożek, J.; Dai, W. Sensory armature and stylets of the mouthparts of *Stephanitis nashi* (Hemiptera: Cimicomorpha: Tingidae), their morphology and function. *Micron* **2020**, *132*, 102840. [CrossRef]

55. Wang, Y.; Brożek, J.; Dai, W. Morphological disparity of the mouthparts in polyphagous species of Largidae (Heteroptera: Pentatomomorpha: Pyrrhocoroidea) reveals feeding specialization. *Insects* **2020**, *11*, 145. [CrossRef] [PubMed]

56. Finkelstein, R.; Perrimon, N. The molecular genetics of head development in *Drosophila melanogaster*. *Development* **1991**, *112*, 899–912. [PubMed]

57. Popadić, A.; Panganiban, G.; Rusch, D.; Shear, W.A.; Kaufman, T.C. Molecular evidence for the gnathobasic derivation of arthropod mandibles and for the appendicular origin of the labrum and other structures. *Dev. Genes Evol.* **1998**, *208*, 142–150. [CrossRef] [PubMed]

58. Boyan, G.; Williams, J.; Posser, S.; Bräunig, P. Morphological and molecular data argue for the labrum being non-apical, articulated, and the appendage of the intercalary segment in the locust. *Arthropod Struct. Dev.* **2002**, *31*, 65–76. [CrossRef]

59. Spooner, C.S. The phylogeny of the Hemiptera based on a study of the head capsule. *Illinois Biol. Monogr.* **1938**, *16*, 1–102. [CrossRef]

60. Gupta, A.P. A consideration of the systematic position of the Saldidae and the Mesoveliidae (Hemiptera: Heteroptera). *Proc. Entomol. Soc. Wash.* **1963**, *65*, 31–38.

61. Štys, P. On the morphology of the labrum in Heteroptera. *Acta Entomol. Bohemoslov.* **1969**, *66*, 150–158.

62. Forthman, M.; Weirauch, C. Phylogenetics and biogeography of the endemic Madagascan millipede assassin bugs (Hemiptera: Reduviidae: Ectrichodiinae). *Mol. Ph. Evol.* **2016**, *100*, 219–233. [CrossRef]

63. Forthman, M.; Weirauch, C. Millipede assassins and allies (Heteroptera: Reduviidae: Ectrichodiinae, Tribelocephalinae): Total evidence phylogeny, revised classification and evolution of sexual dimorphism. *Syst. Èntomol.* **2017**, *42*, 575–595. [CrossRef]

64. Weber, H. Zur vergleichenden Physiologie der Saugorgane der Hemipteren. *Zeitschrift für vergleichende Physiologie* **1928**, *8*, 145–186. [CrossRef]

65. Snodgrass, R.E. *Principles of Insect Morphology*; McGraw Hill Book Co.: New York, NY, USA, 1935.

66. Esquivel, J.F. Stylet penetration estimates for a suite of phytophagous hemipteran pests of row crops. *Environ. Èntomol.* **2015**, *44*, 619–626. [CrossRef] [PubMed]

67. Faucheux, M.M. Relations entre l'ultrastructure des stylets manibulaires et maxillaires et la prise de nourriture chez les insects Hemipteres. *CR Acad. Sci. Paris (Ser. D)* **1975**, *281*, 41–44.

68. Cohen, A.C. Feeding adaptations of some predaceous Hemiptera. *Ann. Èntomol. Soc. Am.* **1990**, *83*, 1215–1223. [CrossRef]

69. Swart, C.C.; Felgenhauer, B.E. Structure and function of the mouthparts and salivary gland complex of the giant Waterbug, *Belostoma lutarium* (Stål) (Hemiptera: Belostomatidae). *Ann. Èntomol. Soc. Am.* **2003**, *96*, 870–882. [CrossRef]

70. Roitberg, B.D.; Gillespie, D.R.; Quiring, D.M.J.; Alma, C.R.; Jenner, W.H.; Perry, J.; Peterson, J.H.; Salomon, M.; VanLaerhoven, S. The cost of being an omnivore: Mandible wear from plant feeding in a true bug. *Naturwissenschaften* **2005**, *92*, 431–434. [CrossRef]

71. Romani, R.; Salerno, G.; Frati, F.; Conti, E.; Isidoro, N.; Bin, F. Oviposition behaviour in *Lygus rugulipennis*: A morpho-functional study. *Èntomol. Exp. Appl.* **2005**, *115*, 17–25. [CrossRef]

72. Bérenger, J.; Pluot-Sigwalt, D. Notes sur *Micrauchenus lineola* (Fabricius 1787), espÈcetermitophile et termitophage (Heteroptera: Reduviidae: Harpactorinae, Apiomerini). *Ann. Soc. Entomol. Fr.* **2009**, *45*, 129–133. [CrossRef]

73. Sahayaraj, K.; Kanna, A.V.; Kumar, S.M. Gross morphology of feeding canal, salivary apparatus and digestive enzymes of salivary gland of *Catamirus brevipennis* (Serville) (Hemiptera: Reduviidae). *J. Entomol. Res. Soc.* **2010**, *12*, 37–50.

74. Brożek, J. A comparison of external and internal maxilla and mandible morphology of water bugs (Hemiptera: Heteroptera: Nepomorpha). *Zootaxa* **2013**, *3635*, 340–378. [CrossRef]

75. Stubbins, F.L.; Mitchell, P.L.; Turnbull, M.W.; Reay-Jones, F.P.F.; Greene, J.K. Mouthpart morphology and feeding behavior of the invasive kudzu bug, *Megacopta cribraria* (Hemiptera: Plataspidae). *Invertebr. Boil.* **2017**, *136*, 309–320. [CrossRef]

76. Cohen, A.C. Feeding fitness and quality of domesticated and feral predators: Effects of long-term rearing on artificial diet. *Boil. Control.* **2000**, *17*, 50–54. [CrossRef]

77. Miles, P.W. Interaction of plant phenols and salivary phenolases in the relationship between plants and hemiptera. *Èntomol. Exp. Appl.* **1969**, *12*, 736–744. [CrossRef]

78. Miles, P. Contact chemoreception in some Heteroptera, including chemoreception internal to the stylet food canal. *J. Insect Physiol.* **1958**, *2*, 338–347. [CrossRef]

79. Backus, E.A.; McLean, D.L. The sensory systems and feeding behavior of leafhoppers. I. The aster leafhopper, *Macrosteles fascifronsstål* (homoptera, cicadellidae). *J. Morphol.* **1982**, *172*, 361–379. [CrossRef] [PubMed]

80. Foster, S.; Goodman, L.J.; Duckett, J.G. Ultrastructure of sensory receptors on the labium of the rice brown planthopper. *Cell Tissue Res.* **1983**, *230*, 353–366. [CrossRef]

81. Frazier, J.L. Nervous system: Sensory system. In *Fundamentals of Insect Physiology*; Blum, M.S., Ed.; John Wiley & Sons: New York, NY, USA, 1985; pp. 287–356.

82. Peregrine, D. Fine structure of sensilla basiconica on the labium of the cotton stainer, *Dysdercus fasciatus* (signoret) (Heteroptera: Pyrrhocoridae). *Int. J. Insect Morphol. Embryol.* **1972**, *1*, 241–251. [CrossRef]

83. Hatfield, L.D.; Frazier, J.L. Ultrastructure of the labial tip sensilla of the tarnished plant bug, *Lygus lineolaris* (P. de Beauvois) (Hemiptera: Miridae). *Int. J. Insect Morphol. Embryol.* **1980**, *9*, 59–66. [CrossRef]

 *insects*

Article

Unique Fine Morphology of Mouthparts in *Haematoloecha nigrorufa* (Stål) (Hemiptera: Reduviidae) Adapted to Millipede Feeding

Yan Wang [1,†], Junru Zhang [1,†], Wanshan Wang [1], Jolanta Brożek [2] and Wu Dai [1,*] 

[1] Key Laboratory of Plant Protection Resources and Pest Integrated Management of the Ministry of Education, College of Plant Protection, Northwest A&F University, Yangling 712100, Shaanxi, China; wangyan105422@163.com (Y.W.); millipedeassassin@gmail.com (J.Z.); wangwanshang1004@163.com (W.W.)

[2] Faculty of Natural Science, Institute of Biology, Biotechnology and Environmental Protection, University of Silesia in Katowice, Bankowa 9, 40-007 Katowice, Poland; jolanta.brozek@us.edu.pl

* Correspondence: daiwu@nwsuaf.edu.cn; Tel.: +89-29-8708-2098

† These authors contributed equally to this work.

Received: 31 May 2020; Accepted: 19 June 2020; Published: 22 June 2020

Abstract: Millipede assassin bugs are a diverse group of specialized millipede predators. However, the feeding behavior of Ectrichodiinae remains poorly known, especially how the mouthpart structures relate to various functions in feeding. In this study, fine morphology of the mouthparts and feeding performance of *Haematoloecha nigrorufa* (Stål, 1867) was observed and described in detail for the first time. The triangular labrum is divided by a conspicuous transverse membrane into a strongly sclerotized basilabrum and a less sclerotized distilabrum. Fifteen types of sensilla are distributed on the mouthparts. Each mandibular stylet has an expanded spatulate apex and about 150 approximately transverse ridges on the external middle side; these help in penetrating the ventral trunk area and the intersegmental membranes of millipede prey. The right maxilla is tapered. On the internal surface are a row dorsal short bristles near the apex and a row of ventral bristles preapically. A longitudinal row of long lamellate structures extend proximate for a considerable distance, lie entirely within the food canal, and bear several short spines and short bristles. There is no obvious difference between males and females in the distribution, number, and types of sensilla on mouthparts. The adult feeding process involves several steps, including searching and capturing prey, paralyzing prey, a resting phase, and a feeding phase. The evolution of the mouthpart morphology and the putative functional significance of their sensilla are discussed, providing insight into the structure and function of the mouthparts adapted for millipede feeding.

Keywords: Reduviidae; mouthparts; sensillum; feeding; predation

1. Introduction

Mouthparts are the feeding organs of insects [1–3] and morphological variation in mouthparts generally corresponds well to their different feeding requirements [4,5]. A large number and variety of sensilla are attached to the different mouthparts, which play important roles in host search, detection, feeding, and mating [6–9]. The types and quantities of various sensilla are closely related to the feeding habits of insects. The mouthpart complex in Hemiptera is often called the "rostrum", "sucking beak", or piercing-sucking mouthparts [10]. During the long-term evolutionary history of Hemiptera, the components of this feeding apparatus have been clearly modified to serve their unique functions in different groups of bugs [11–16] and specialized to various sources of food [17]. Heteroptera, as compared to other groups of hemipterans (Cicadomorpha, Fulgoromorpha, Sternorrhyncha, and Colleorrhyncha) display a wider array of trophic and morphological diversity [12].

The sensillar systems of heteropteran mouthparts have been studied in representatives across different trophic groups and different families often display specific characteristics under examination [15,16,18–25]. Although data on some aspects of mouthpart morphology of Heteroptera are abundant, based on light and scanning electron microscopy, previous observations have been reported at various levels of accuracy [12,26–36]. Thus, consistent and detailed studies are needed in order to provide useful comparative data.

The feeding behavior of Hemipterans has been inferred from mouthpart structures [31,32], or was based on electropenetrograph (EPG) apparatus [37,38]. Unfortunately, ethological details about feeding behavior are lacking [39–41]. The fine morphology of mouthparts allows for us to interpret the function of the component parts of the feeding apparatus and provides us with information to understand the actual feeding mechanism.

Assassin bugs, or Reduviidae (Insecta: Heteroptera), are the second largest family and one of the most morphologically diverse groups of true bugs, with more than 6600 species [42]. Some data suggest that the evolutionary transition to predation in this family coincided with the reduction of one segment of their labium [43]. Generally, the reduviid rostrum is three-segmented, short and thick, usually curved and arched, and, when at rest, the labium is held in a longitudinal groove (the friction groove) in the center of the anterior thoracic and abdominal plates.

Details of mouthpart morphology of reduviid species have received only sporadic. Previous studies have mostly concentrated on the terminal labial sensilla [12,18,44], gross morphology of the mandible and maxilla [45,46], interlocking mechanisms of maxillae and mandibles [47], and gross morphology of the labium and labrum [46,48–52]. More detailed information on fine structure is needed to determine how much variability in mouthparts occurs across different taxa, particularly those with relatively narrow feeding specialization.

Millipede assassin bugs (Hemiptera: Reduviidae: Ectrichodiinae) are a diverse group with more than 700 species known worldwide [52]. Many ectrichodiines appear to be specialized millipede (Diplopoda) predators, but details of the predator-prey relationships, including prey specificity and point of mouthpart insertion are largely undocumented [53]. *Haematoloecha nigrorufa* (Stål, 1867) is a common and widespread species of assassin bug in China (Beijing, Zhejiang, Sichuan, Fujian, Jiangxi, Jiangsu, Gansu, Liaoning, Shaanxi) and some other Asian countries (Japan, Korean Peninsula), mainly feeding on millipedes and some arthropod pests. The fine structure of the mouthparts of *H. nigrorufa*, and the significance of these mouthpart structures for feeding, have not been previously studied.

This paper describes morphological observations of the mouthparts and feeding performance of *H. nigrorufa* to expand our knowledge of modifications of these organs and their sensilla system in a species that is adapted to the difficulties of obtaining food from millipedes. We focused on sensilla typology, distribution, and possible functions with the aim of identifying more character sets that are useful for future comparative morphological studies in Reduviidae.

2. Material and Methods

2.1. Insect Collecting

Adults of *H. nigrorufa* used for SEM in this study were collected with sweep nets at the campus of Northwest A&F University in Yangling, Shaanxi Province, China (34°160 *n*, 108°070 E, elev. 563 m) in August 2019, preserved in 75% ethanol, and stored at 4°C. For observing the performance of the mouthparts during feeding inside different types of prey, additional adults of *H. nigrorufa* were collected at the same locality in September 2019. Prey animals were not dissected after feeding.

2.2. Samples for SEM

Adult males ($n = 4$) and females ($n = 12$) were dipped into 10% NaOH solution for 2h and cleaned twice while using an ultrasonic cleaner (KQ118, Kunshan, China) 15s each time. Dehydration used serial baths of 80%, 90% and 100% ethanol each for 15 min. The materials were air dried, coated with a film of gold (Q150T-S, Quorum, West Sussex, UK), and then imaged with a Nova Nano SEM-450 (FEI, Hillsboro, OR, USA) at 5–10 kV in the scanning microscopy laboratories of the Life Science Research Core Services of Northwest A & F University.

2.3. Feeding Behavior on Millipedes

The observation of predatory behavior in *H. nigrorufa* lasted for two months and insects were bred in a transparent plastic basin 150 mm diameter, 50 mm tall. The bottom of the basin was covered with cotton wool 15 mm high, which was moistened with about 50 ml of water. On the cotton wool, bark and some leaves were placed to simulate the natural environment. The feeding conditions were (24 ± 1) °C, RH (70 ± 15) %. Mealworms, armyworms, earwigs, polydesmid millipedes, juliform millipedes, and spirostreptid millipedes were separately used as prey. After 72 hours of fasting, the prey was offered to 20 adult individuals of *H. nigrorufa* in the basin and observations of the results. The predatory behavior was captured by a Nikon D500 camera and the images were imported into a computer for later analysis.

2.4. Image Processing and Morphometric Measurement

Photographs and SEMs were observed and measured after being imported into Adobe Photoshop CC 2019 (Adobe Systems, San Jose, CA, USA).

2.5. Data Analysis

The lengths of the mouthpart were compared between sexes using Student t-test. Statistical analyses were executed using SPSS 19.0 (SPSS, Chicago, IL, USA).

2.6. Terminology

The sensilla were classified according to their external morphology, length, distribution, and position. The terminology of sensilla follows Altner and Prillinger [54] and Frazier [55], with less specialized nomenclature from Brożek and Chłond [18]. Terminology of the labium and stylet bundle structures follow Weirauch [42,43], Cobben [12], Brożek, and Herczek [47], with some new terms established based on the present study.

3. Results

3.1. General Morphology of Mouthparts

The main elements of the mouthparts of *H. nigrorufa* are similar to other reduviid species and include the two-part labrum, three-segmented labium (the order of segment numbers in this group is from II–IV) (Figure 1A–C), and stylet fascicle (Sf) composed by two separated mandibular stylets (Md) and two interlocked maxillary stylets (Mx) (Figure 1B). No obvious differences were noted between the mouthpart structure of females and males except for their length ($t(13) = 2.235$, $p = 0.044$) (Table 1).

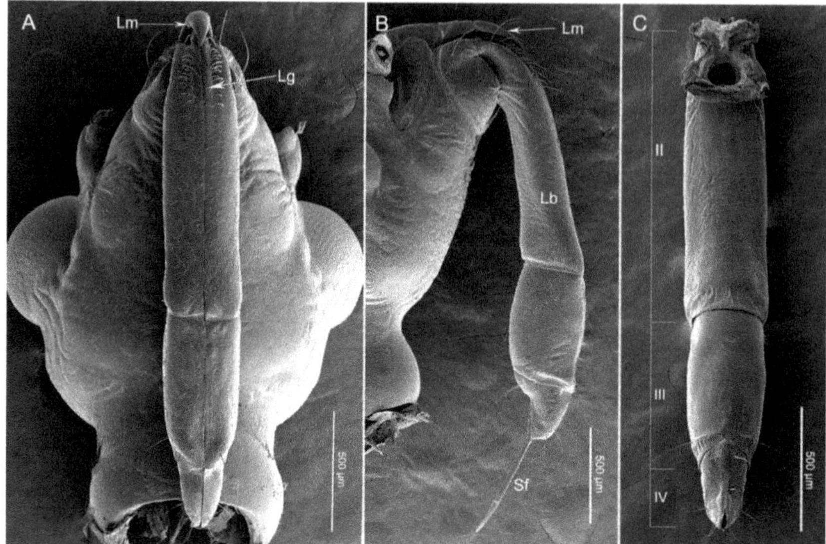

Figure 1. Scanning electron micrographs of the head of *Haematoloecha nigrorufa*. (**A**). Ventral view; (**B**). Lateral view; (**C**). Dorsal view showing three-segmented labium (II–IV); Sf, stylet fascicle; Lm, labrum; Lb, labium; Lg, labial groove.

Table 1. Measurements of labrum, labium and stylets (mean ± SE) obtained from scanning electron microscopy. Lb, labium; Lb, 2, 3, 4, the second, third, fourth segment of labium; Lm, labrum; Md, mandibular stylets; RMx, right maxillary stylet; LMx, left maxillary stylet; N = sample number.

Sex	Position	Length (μm)	Width (μm)	N
Female	Lb	2058.0 ± 25.8	-	12
	L2	1174.8 ± 35.9	215.3 ± 4.4	12
	L3	587.4 ± 15.7	301.7 ± 5.8	12
	L4	298 ± 18.3	194 ± 2.8	12
Male	Lm	476.5 ± 8	-	4
	Lb	1875.2 ± 113.5	-	3
	Lb2	1011 ± 131.2	219 ± 8.7	3
	Lb3	583.6 ± 14.2	288.2 ± 1.2	3
	Lb4	285.8 ± 18.9	183 ± 4.4	3
	Md (spatulate apex)	507.4 ± 15.3	-	4
	RMx (dorsal short bristles)	104.3 ± 6.5	-	3
	RMx (ventral lamellate structures)	424.2 ± 13.7	-	3
	LMx (dorsal short spines)	15.8 ± 0.5	-	3
	LMx (ventral short bristles)	76.4 ± 3.6	-	3

3.2. Labrum

The triangular labrum reaches the base of the second labial segment (Figure 1A,B and Figure 2A). The labrum is divided by a conspicuous transverse membranous zone into two unequally sclerotized parts: a strongly sclerotized basilabrum (bl) and a less sclerotized distilabrum (dl) (Figure 2A,C,D). The basilabrum is wide and it accounts for 1/3 of the length of the labrum. The surface of the basilabrum is slightly plicated and sensilla trichodea (St1) and multilobular sensilla (Sm) are sparsely distributed on it (Figure 2A,C). The distilabrum (dl) is elongated and cone-shaped. In this part of the labrum, sensilla trichodea (St1) and multilobular sensilla (Sm) are less numerous. Moreover, several small spikes (smi) were observed at the tip of the distilabrum (dl) (Figure 2E,H). On the ventral side of the distilabrum there is a groove (gr) that holds a proximal part of the stylet fascicle (Sf) (Figure 2D).

Spike-like microtrichia (smi) are irregularly distributed on the ventral surface (Figure 2D,H). Sensilla trichodea I (St1) are at mid-length, slightly curved, and lay flat on the surface of the labrum (Table 2, Figure 2B). The base of the sensilla is set in a pit, the surface has many pores and the tip is rounded (Figure 2B,F). Multilobular sensilla (Sm) are very small, but numerous, placed in cuticular cavities and resemble spread fingers (Table 2, Figure 2B,G).

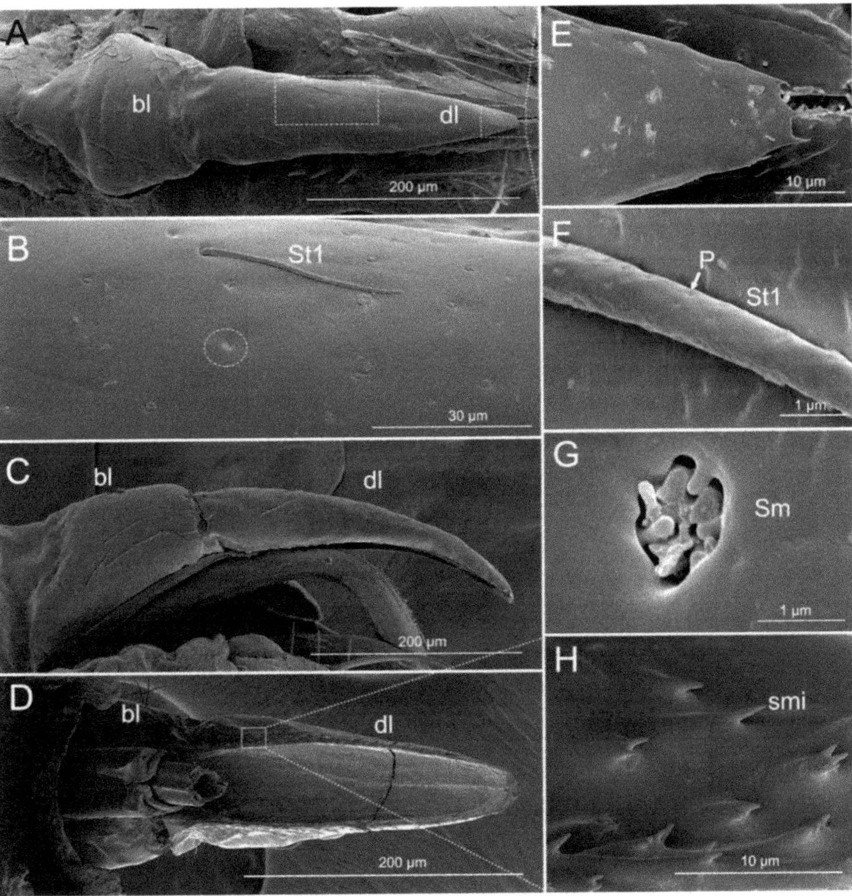

Figure 2. SEM of labrum of *Haematoloecha nigrorufa*. (**A**). Ventral view; (**B**). Enlarged view of box in (**A**), showing sensilla trichodea (St1) and multilobular sensilla (Sm); (**C**). Lateral view; (**D**). Dorsal view; (**E**). Enlarged view of tip of labrum; (**F**). Enlarged view of surface of sensilla trichodea (St1) showing pores (*p*); (**G**). Enlarged view of multilobular sensilla (Sm); (**H**). Enlarged view of box in (**D**), showing spike-like microtrichia (smi); bl, basilabrum; dl, distilabrum.

Table 2. Distribution, morphometric data (mean ± SE), terminology and definition of sensilla used in the present paper after data of prior authors [18,48,49]. Lb2, 3, 4, the second, third, fourth segment of labium; Lm, labrum; N = sample size; St1–4, sensilla trichodea I-IV; Sb1–6, sensilla basiconica I-VII; Sca1–3, sensilla campaniformia; Spe,placoid elongated sensilla; Sm, multilobular sensilla; SF, sensory field on the labial tip; Wp, wall pore; Tp, tip pore.

Type	Distribution and Number	Length (µm)	Basal Diameter (µm)	Shape	Socket	Surface	Pore	Category	Function	N
St1	Lm, Lb2-Lb4	42.8 ± 1.7	1.45 ± 0.1	Hair in pit	Fnflexible	Smooth	Wp (Multiporous)	Chemoreceptive sensilla	Olfactory	20
St2	Lb2 (1 pair), Lb3 (2pairs), Lb4 (5pairs)	142.8 ± 6	5.3 ± 0.2	Hair, peg	Flexible	Grooved	No	Mechanoreceptive sensilla	Tactile	11
St3	Lb2 (about 16 pairs)	103.8 ± 5.8	5.2 ± 0.1	Hair	Flexible	Grooved	No	Mechanoreceptive sensilla	Tactile	10
St4	Lb3 (6 pairs)	9.4 ± 0.6	1.6 ± 0.1	Hair, peg	Flexible	Smooth	No	Mechanoreceptive sensilla	Tactile	14
Sb1	3 pairs at the base of the second segment, 1 pair on the junction between the third and fourth segment	32.7 ± 2.5	4.9 ± 0.1	Peg	Flexible	Smooth	Wp (Uniporous)	Proprioceptive sensilla	Perceive the degree of flexion of the joint	10
Sb2	Lb2 (2 pairs)	9.5 ± 0.7	1.4 ± 0.1	Peg in pit	Inflexible	Smooth	No	Thermo-hygroreceptive sensilla	Temperature/humidity	6
Sb3	Lb4 (4 pairs)	6.0 ± 0.4	1.1 ± 0.1	Peg	Flexible	Smooth	No	Mechanoreceptive sensilla	Tactile	14
Sb4	SF (2 pairs)	5.1 ± 0.4	4.2 ± 0.2	Peg (sensillum coleoconicum)	Inflexible	Smooth	Tp	Chemoreceptive sensilla	Gustatory	8
Sb5	SF (1 pair)	1.3 ± 0.1	1.2 ± 0.1	Peg	Flexible	Smooth	Tp	Chemoreceptive sensilla	Gustatory	4
Sb6	Tip (1 pair)	1.7 ± 0.1	1.2 ± 0.1	Peg in pit	Flexible	Smooth	Tp	Chemoreceptive sensilla	Gustatory	4
Sca1	Lb2-Lb4	-	7.2 ± 0.7	Oval plate	Inflexible	Smooth	Tp	Proprioceptive sensilla	Perceive the degree of flexion of the joint	8
Sca2	SF (1 pair)	-	5.5 ± 0.3	Dome-like structures	Inflexible	Smooth	Tp	Proprioceptive sensilla	Perceive the degree of flexion of the joint	4
Sca3	SF (1 pair)	-	1.2 ± 0.1	Oval plate	Inflexible	Smooth	Tp	Proprioceptive sensilla	Perceive the degree of flexion of the joint	4
Spe	SF (1 pair)	16.8 ± 2.1	2.2 ± 0.1	Dome-elongated	Inflexible	Smooth	Tp	Chemoreceptive sensilla	Gustatory	8
Sm	Lm, Lb2-Lb4	-	1.5 ± 0.1	Pegs in cavity surrounded by fingerlike structures	Inflexible	Smooth	No	Thermo-hygroreceptive sensilla	Temperature/humidity	20

3.3. Labium

The assassin bug *H. nigrorufa* is similar to most other reduviids in having a labium with only three segments. The short and stout three-segmented (II–IV) labium encloses the two mandibular stylets (Md) and two maxillary stylets (Mx) in the labial groove (Lg) (Figure 1B). The groove (Lg) is a shallow depression that is situated on the anterior side of segments two to four. The edges of the groove are close together and form one straight suture above the stylets (Figure 1A). The labial segments differ in morphology and size (Table 1, Figure 1A–C).

Following previous authors we interpret the first labial segment (I) to be either lost or fused to the head capsule. The first visible segment is the second segment (II) and it is the longest segment of the labium (Table 2, Figure 1A–C). The proximal part is slightly narrowed and abundant sensilla are concentrated on this area (Figure 3A) whereas, the distal part of the segment is slightly widened (Figure 3A–C). Seven types of sensilla are found on this segment, including three types of sensilla trichodea (St1, St2, and St3), two types of sensilla basiconia (Sb1 and Sb2), one type of sensilla campaniformia (Sca1), and one type of multilobular sensilla (Sm). Numerous sensilla trichodea (St1) and multilobular sensilla (Sm) are distributed all over the labium II surface (dorsal, lateral, and ventral surfaces). Sensilla trichodea (St1) are distinguished by the presence of many surface pores (Figure 2B,F). One pair of sensilla trichodea (St2) is arranged on the base of segment II. St2 are very long, straight, and almost perpendicular to the surface of the labium (Table 2, Figure 3D,E). The base of the sensillum has a flexible socket, the surface has a vertical groove, and the tip is narrow (Figure 3D,E). Approximately 16 pairs of sensilla trichodea (St3) cover 1/4 of the area of segment II. St3 are long, slightly curved and almost lay flat on the surface of the labium (Table 2, Figure 3D,G). There are three pairs of sensilla basiconica (Sb1) at the base of the second segment (Figure 3D). Sb1 are at middle length, are straight, with a smooth surface, have a base wall pore, a rounded tip, and a flexible socket (Table 2, Figure 3H). Sensilla basiconica (Sb2) are short, small, with a smooth surface, and they have a rounded tip that sits in a pit (Figure 3I). This type of sensillum is sparsely distributed on the ventral surface of the second segment. Sensilla campaniformia (Sca1) are flat, oval-shaped discs with a terminal pore, sparsely distributed on the ventral surface (Table 2, Figure 3F,J).

Labial segment III is straight (Figure 4A–C). The latero-dorsal side of the segment is more expanded in the proximal part than in the distal part. The shorter membrane between segments III and IV viewed from the lateral and dorsal side is usually more or less undulate. Four types of sensilla were found on this segment, including two types of sensilla trichodea (St1 and St2), one type of sensilla campaniformia (Sca1), and one type of multilobular sensilla (Sm). Numerous St1 and Sm are distributed all over the segment III surface (dorsal, lateral, and ventral surfaces) (Figure 4A–D). One pair of St2 is found on the distal part of the ventral surface and one pair is found laterally (Figure 4D). One pair of Sca1 is distributed on the dorsal surface (Figure 4E).

The labial segment IV is short, blunt, and robust (Figure 5A–C). The shape of the segment is slightly conical, because the ventral side is somewhat convex, in contrast to the lateral–dorsal surface (Figure 5B) with a distinct concavity covering 1/3 of the area. The proximal part of the segment is almost as wide as the end of the third segment (Figure 5B). The end of segment IV is narrowed and slightly bent ventrad, appearing hooklike. A stridulatory organ is composed of two parts, namely a plectrum on the end of the fourth labial segment and a stridulitrum on the prothorax, is present. In the analyzed specimens, the plectrum consisted of one sclerotized tubercle (tr) on each of the paired lateral lobes of the apex of the last labial segment (Figure 5E).

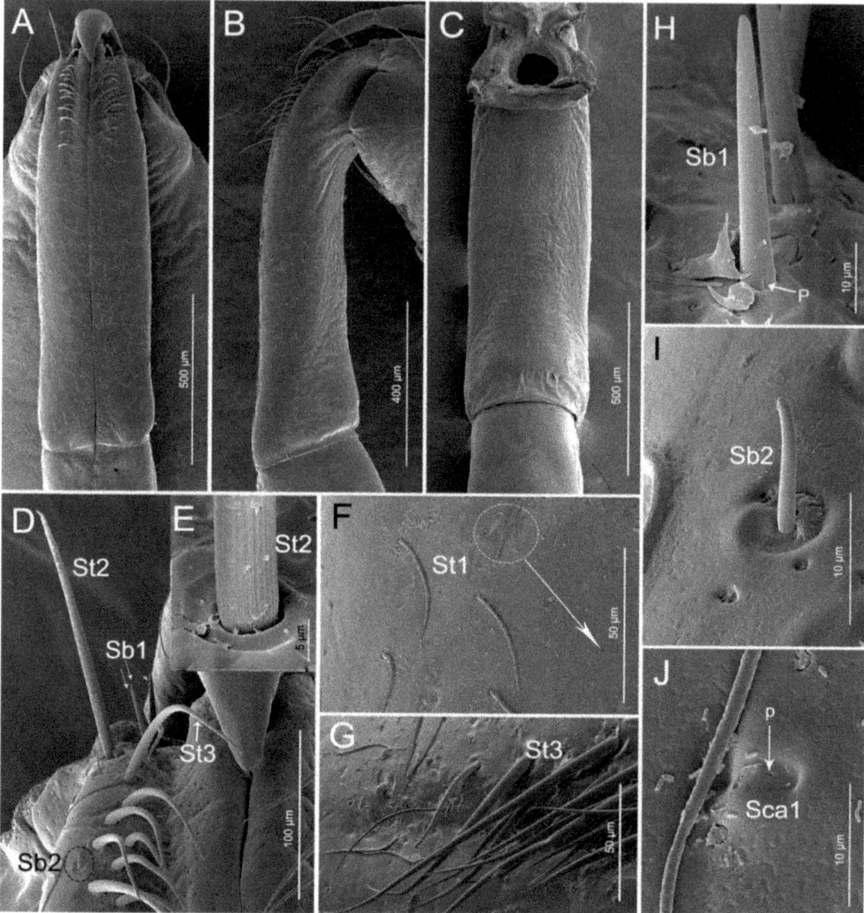

Figure 3. SEM of second labial segment of *Haematoloecha nigrorufa*. (**A**). Ventral view; (**B**). Lateral view; (**C**). Dorsal view; (**D**). Base of the second segment showing sensilla trichodea (St2), sensilla trichodea (St3), sensilla basiconica (Sb1) and sensilla basiconica (Sb2); (**E**). Base view of sensillum trichodeum (St2); (**F**). Enlarged view of dorsal surface of labium showing sensilla campaniformia (Sca1) and sensilla trichodea (St1); (**G**). Sensilla trichodea (St3); (**H**). Sensilla basiconica (Sb1); (**I**). Sensilla basiconica (Sb2); (**J**). Sensilla campaniformia (Sca1); *p*, pore.

Thirteen types of sensilla were found on this segment, including three types of sensilla trichodea (St1, St2, and St4), five types of sensilla basiconica (Sb1, Sb3, Sb4, Sb5, and Sb6), three types of sensilla campaniformia (Sca1, Sca2, and Sca3), one type of sensilla placoid elongated (Spe), and one type of multilobular sensilla (Sm). There are many sensilla trichodea (St1) and multilobular sensilla (Sm) all over the segment IV dorsal, lateral, and ventral surfaces (Figure 5A–H). Five pairs of sensilla trichodea (St2) are found on segment IV (two pairs on the ventral surface, one pair on the lateral surface, and two pairs on the dorsal surface). Six pairs of sensilla trichodea (St4) are distributed on the proximal part of the fourth segment (Figure 5E). One pair of sensilla basiconica (Sb1) is distributed between the third and fourth labial segments (Figure 5A). Four pairs of sensilla basiconica (Sb3) are found on the distal part of the ventral surface (Figure 5D,E). Eight pairs of sensilla campaniformia (Sca1) are distributed on the proximal position of the last segment (one pair on the ventral surface, two pairs on the lateral surface, and five pairs on the dorsal surface) (Figure 5D,E,G, and Figure 6A).

The labial tip has two lateral lobes and the sensilla are symmetrically arranged on this area and form two sensory fields, including sensilla trichodea (St4), sensilla basiconica (Sb4, Sb5, and Sb6), sensilla campaniformia (Sca1, Sca2, and Sca3), and placoid elongated sensilla (Spe) (Figure 6A–J). Hair-like sensilla trichodea (St4) are short, small, with a smooth surface and they have a rounded tip and a flexible socket (Figure 6B,D). Peg-like sensilla basiconia (Sb3) are short, small, with a smooth surface and they have a narrowed tip and a flexible socket (Figure 5H). Sensilla basiconica (Sb4), similar to sensilla coeloconica, are short cones that arise from inflexible sockets. The base of the sensillum with the socket is elevated above the surrounding cuticle. A sensillum may either extend beyond the socket or remain hidden inside the base (Figure 6C). Sensilla basiconica (Sb5) are a short, flattened cone with additional small processes at the end; the cone base is embedded in a flexible socket (Figure 6H). Sensilla basiconica (Sb6) are short cones with a single pore at the tip and they have a flexible socket (Figure 6B,J). Sensilla campaniformia (Sca2) are flat, oval disks with a single pore on their surface (Figure 6F). Placoid elongated sensilla (Spe) are elongated oval plates that have no pores (Figure 6E).

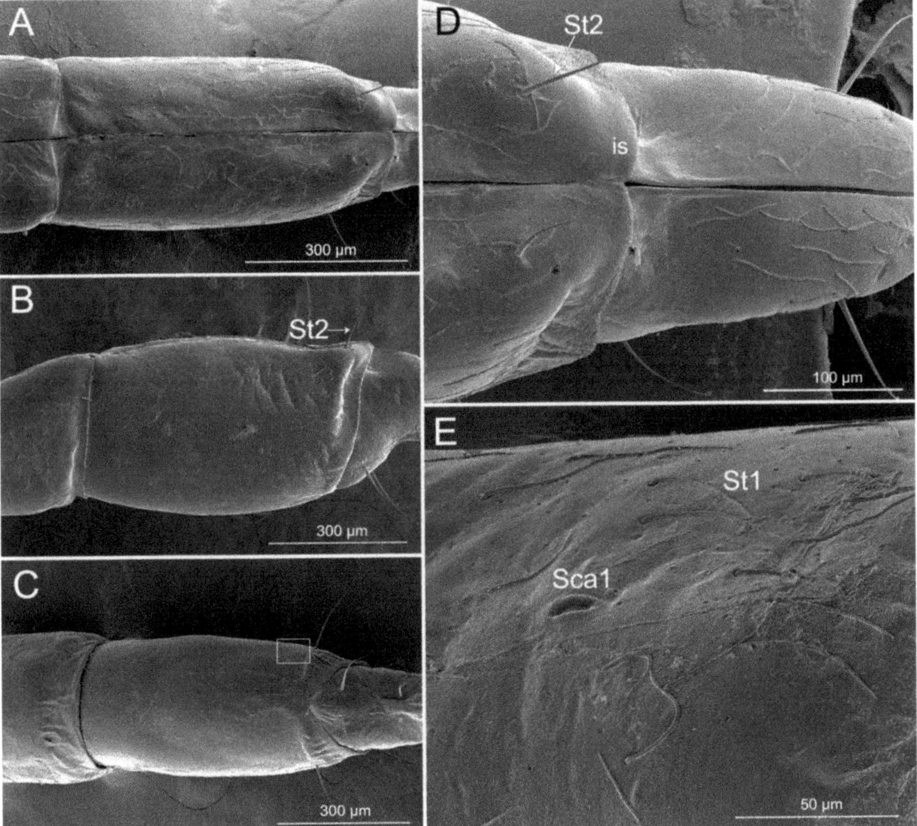

Figure 4. SEM of the third labial segment of *Haematoloecha nigrorufa*. (**A**). Ventral view; (**B**). Lateral view showing sensilla trichodea (St2); (**C**). Dorsal view; (**D**). Enlarged view of dorsal surface showing sensilla trichodea (St2) and intercalary sclerite (is); (**E**). Enlarged view of box in (**C**), showing sensilla campaniformia (Sca1) and sensilla trichodea (St1).

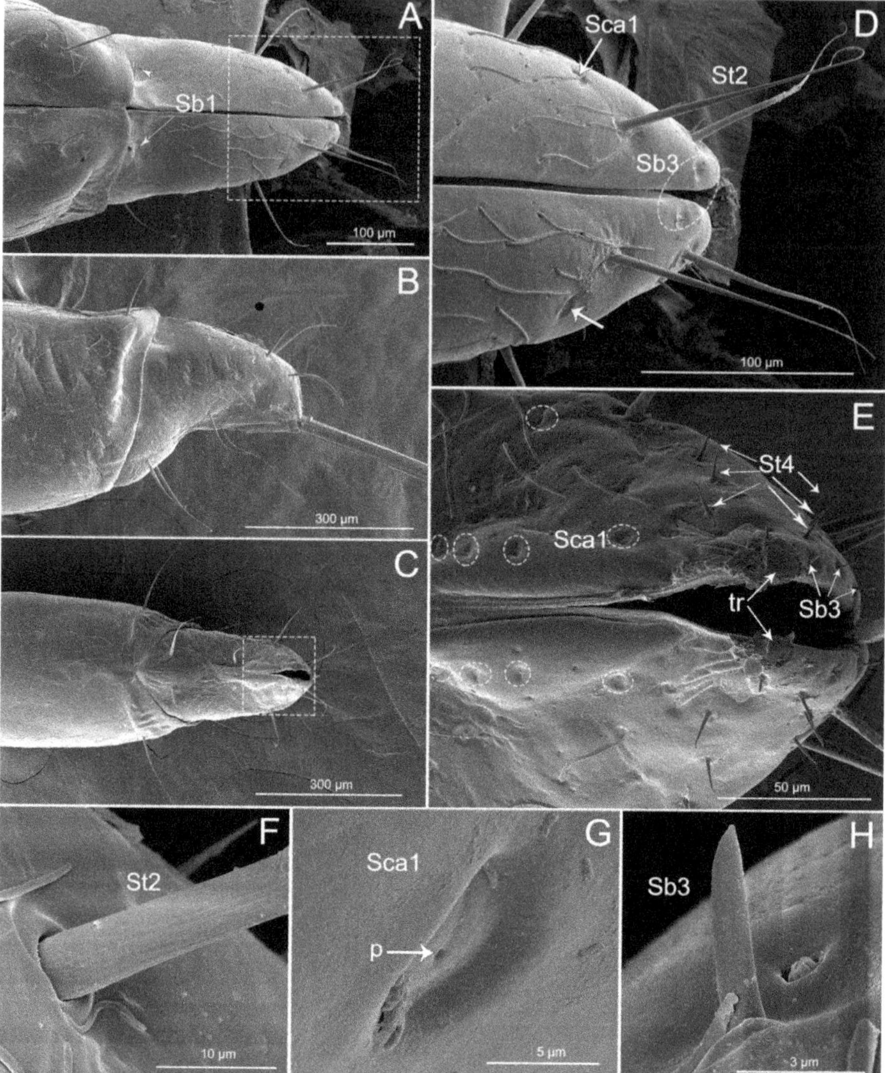

Figure 5. SEM of the fourth labial segment of *Haematoloecha nigrorufa*. (**A**). Ventral view showing a pair of sensilla basiconica (Sb1); (**B**). Lateral view; (**C**). Dorsal view; (**D**). Enlarged view of box in (**A**), showing sensilla trichodea (St2), sensilla campaniformia (Sca1) (white arrows) and sensilla basiconica (Sb3) (white circle); (**E**). Enlarged view of box in (**C**), showing some sensilla campaniformia (Sca1) (white circles), sensilla trichodea (St4) and sensilla basiconica (Sb3); (**F**). Enlarged view of sensillum trichodeum (St2); (**G**). Enlarged view of sensilla campaniformia (Sca1); (**H**). Enlarged view of sensilla basiconica (Sb3) and multilobular sensilla (Sm); *p*, pore.

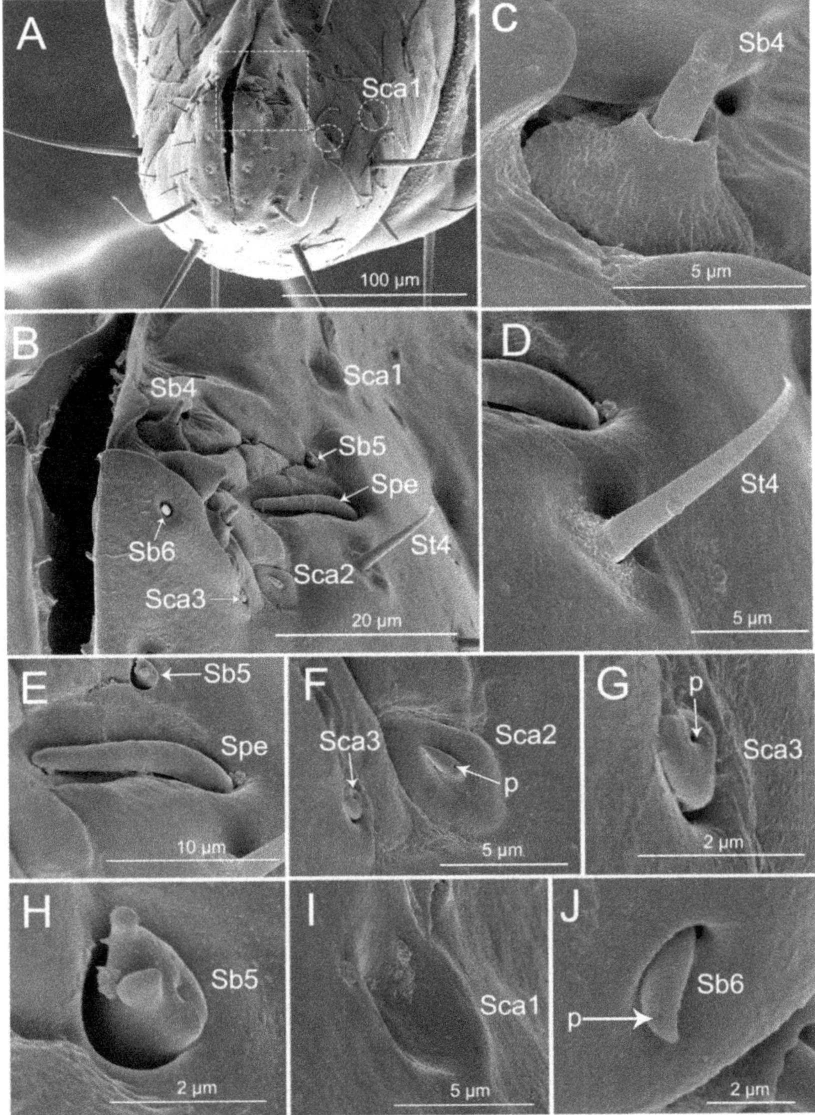

Figure 6. SEM of tip of labium of *Haematoloecha nigrorufa*. (**A**). Vertical view; (**B**). Enlarged view of box in (**A**), showing sensilla trichodea (St4), sensilla placodeum elongated (Spe), three type of sensilla campaniform (Sca1, Sca2, Sca3), and three type of sensilla basiconica (Sb4, Sb5, Sb6); (**C**). Sensilla basiconica (Sb4); (**D**). Sensilla trichodea (St4); (**E**). Placoid elongated sensilla (Spe) and sensilla basiconica (Sb5); (**F**). Sensilla campaniformia (Sca2) and sensilla campaniformia (Sca3); (**G**). Sensilla campaniformia (Sca3); (**H**). Sensilla basiconica (Sb5); (**I**). Sensilla campaniformia (Sca1); (**J**). Sensilla basiconica (Sb6); *p*, pore.

3.4. Stylet Fascicle

The stylet fascicle is composed of two separated mandibular stylets (Md) and two interlocked maxillary stylets (Mx). The mandibular stylets (Md) are slightly shorter than the maxillary stylets (Mx), and right maxillary stylets (RMx) are slightly longer than left maxillary stylets (LMx).

The mandibular stylets (Md) are addressed laterally to the maxillary stylets (Mx). In *H. nigrorufa*, the external side of the Md has a spatulate apex with about 150 slightly transverse ridges (str) (Figure 7A,B) and the end is narrowed. Based on the transverse ridges, there are numerous strong longitudinal ridges (slr) that are extended all the way to the base. On the inner surface of the spatulate apex, dorsally, and ventrally are many visible longitudinal ridges (lr). Between them the surface is smooth. Based of the spatulate apex, there are small spikes (ss) on the middle of the inner surface (Figure 7C,D). In this specie, the left-right asymmetry of the maxillary stylets is noticeable.

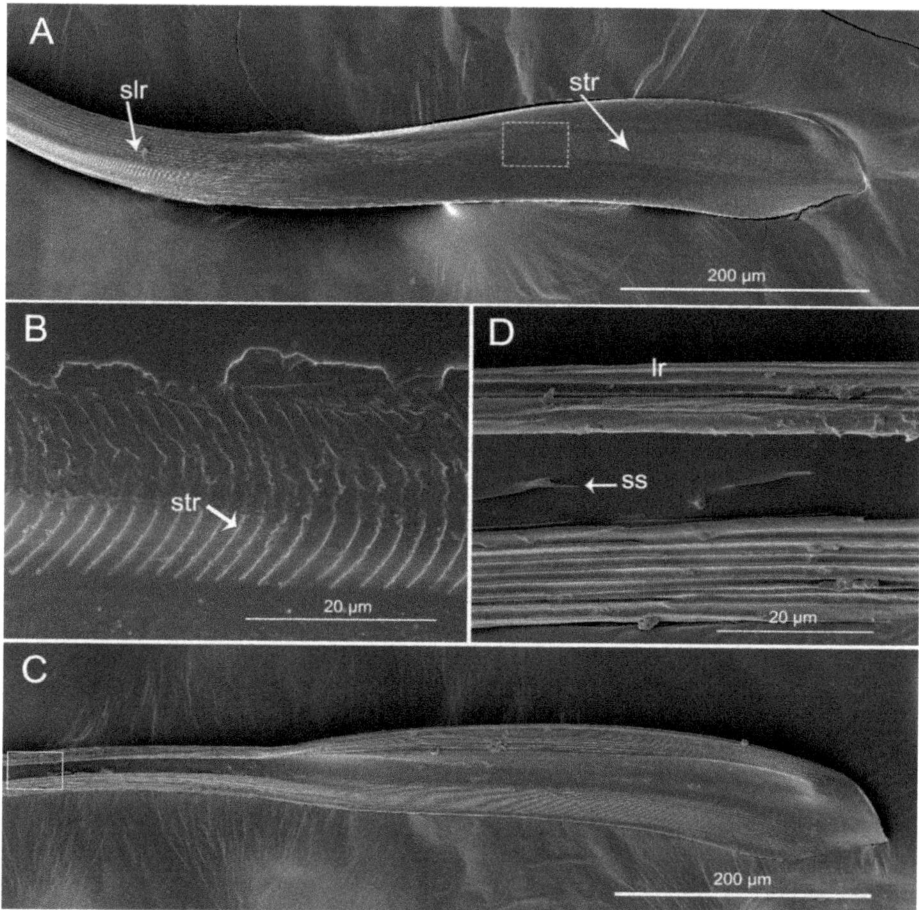

Figure 7. SEM of mandibular stylets of *Haematoloecha nigrorufa*. (**A**). External view; (**B**). Enlarged view of box in (**A**); (**C**). Interior view; (**D**). Enlarged view of box in (**C**); slr, strong longitudinal ridges; str, slightly transverse ridges; lr, longitudinal ridges; ss, small spinule.

The apex of the right maxilla (RMx) is tapered (Figure 8A,C) and it has a distinct curvature (Figure 8D). The external side of the right maxillary stylet apex of the subapical region is smooth (Figure 8G) with ventral (vr) and dorsal rows (dr) of curved hair-like short bristles (sbr), and the ventral row possesses lamellate structures (lss) (Figure 8A–E). On the lateral surface of the right maxillary stylet (RMx) apex there is a submedial row of many small teeth (sto) (Figure 8F) and a lateral band of many small pores (spo) (Figure 8F). The apex of the left maxilla is straight with a notch (no) and a narrow lobe (nlo) directed anteriorly (Figure 9A,E). On the internal surface of the dorsal (dr) side, there are several short spines (ssp) and many small teeth (sto) (Figure 9D,E). On the ventral (vr) side of the left maxilla, there are some transverse ridges (tr) and short bristles (sbr) (Figure 9A,B, and Figure 10A–C). On the external surface of the left maxillary stylet there are many small pores (spo) (Figure 10C–E).

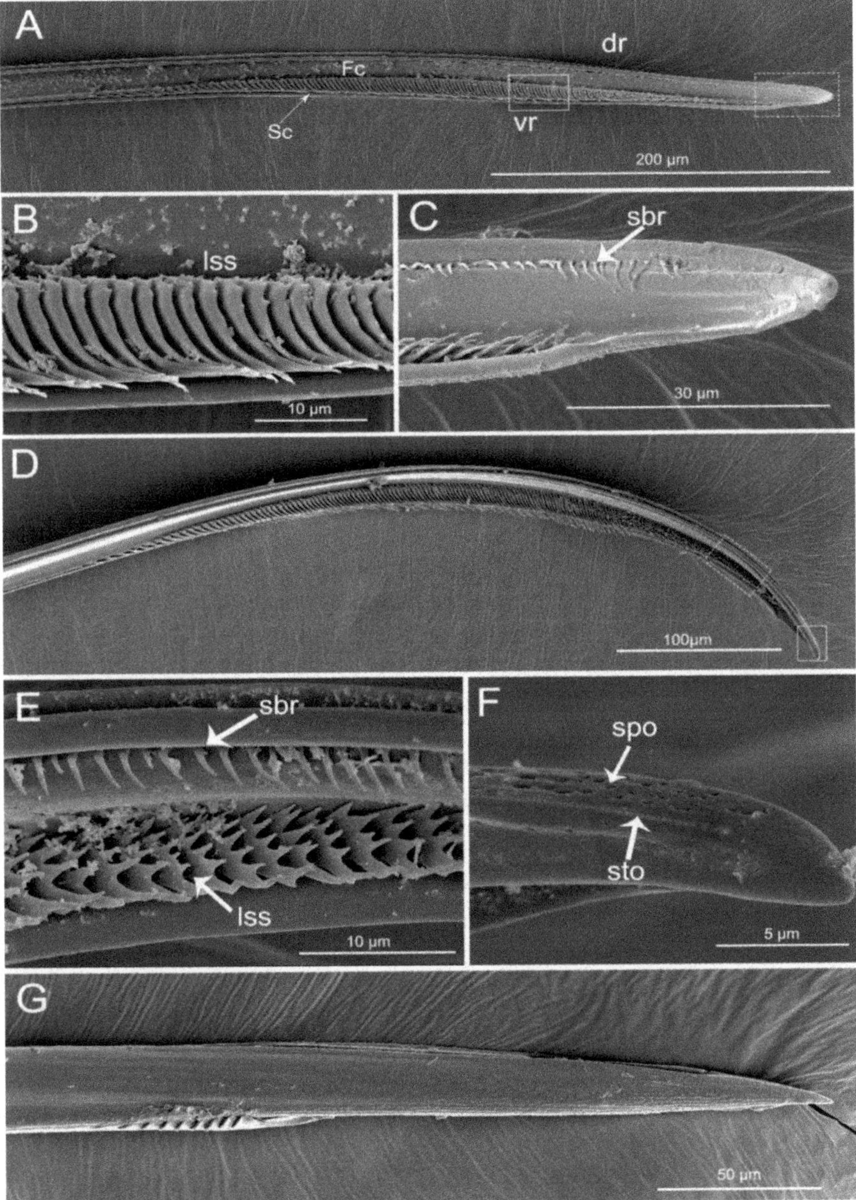

Figure 8. SEM of right maxillary stylets of *Haematoloecha nigrorufa*. (**A**). Interior view; (**B**). Enlarged view of box in (**A**); (**C**). Enlarged view of box in (**A**), showing tip of right maxillary stylet (RMx); (**D**), Lateral view; (**E**). Enlarged view of box in (**D**); (**F**). Enlarged view of box in (**D**), showing tip of right maxillary stylet (Rmx); (**G**). External view; dr, dorsal row; vr, ventral row; lss, lamellate-shaped structures; sbr, short bristles; spo, small pore; sto, small tooth.

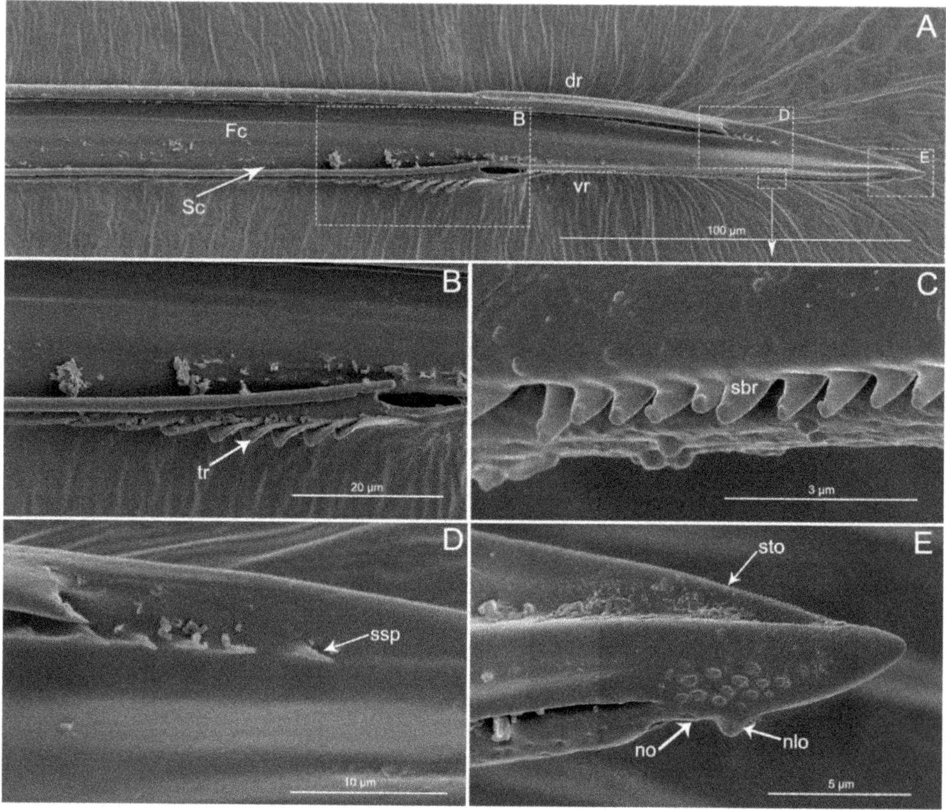

Figure 9. SEM of left maxillary stylets of *Haematoloecha nigrorufa*. (**A**). Interior view; (**B–E**). Enlarged view of box in (**A**); tr, transverse ridges; sbr, short bristles; no, notch; nlo, narrow lobe; sto, small tooth; ssp, short spines.

Cross-sections through the labium and stylet bundle (interlocked maxillae and mandibles) in reduviid species show that the stylet bundles are distinctly laterally compressed (Figure 11A–F). In the cross-section of the labium, the dorsal walls are visible (dw) (edges). The edges are close and they form a tight connection. The floor of the labial groove (lg) is deeper, located under the dorsal edges and envelopes the stylet bundle (Figure 11A,B).

The maxillary stylets are encased by the outer, overlapping mandibular stylets. The shape of the cross-section of the maxillary and mandibular stylets changes from base to end (Figure 11A–F). The two maxillae are held together by interlocking processes forming three locks: dorsal, medial, and ventral. The dorsal lock has two hooked processes and two straight processes. The middle lock has two hooked processes. The ventral lock has one straight and two hooked processes. Within each maxillary stylet there is one axial canal (ac) with three dendrites (de) (Figure 11D). The mandibles are expanded on the ventral side and they possess a wide nerve canal. On the inner ventral wall there is a longitudinal ridge that corresponds to the longitudinal ridges (lr) on Figure 7D. Lateral and dorsal parts of the mandibles are thinner than ventral portions (Figure 11A–D).

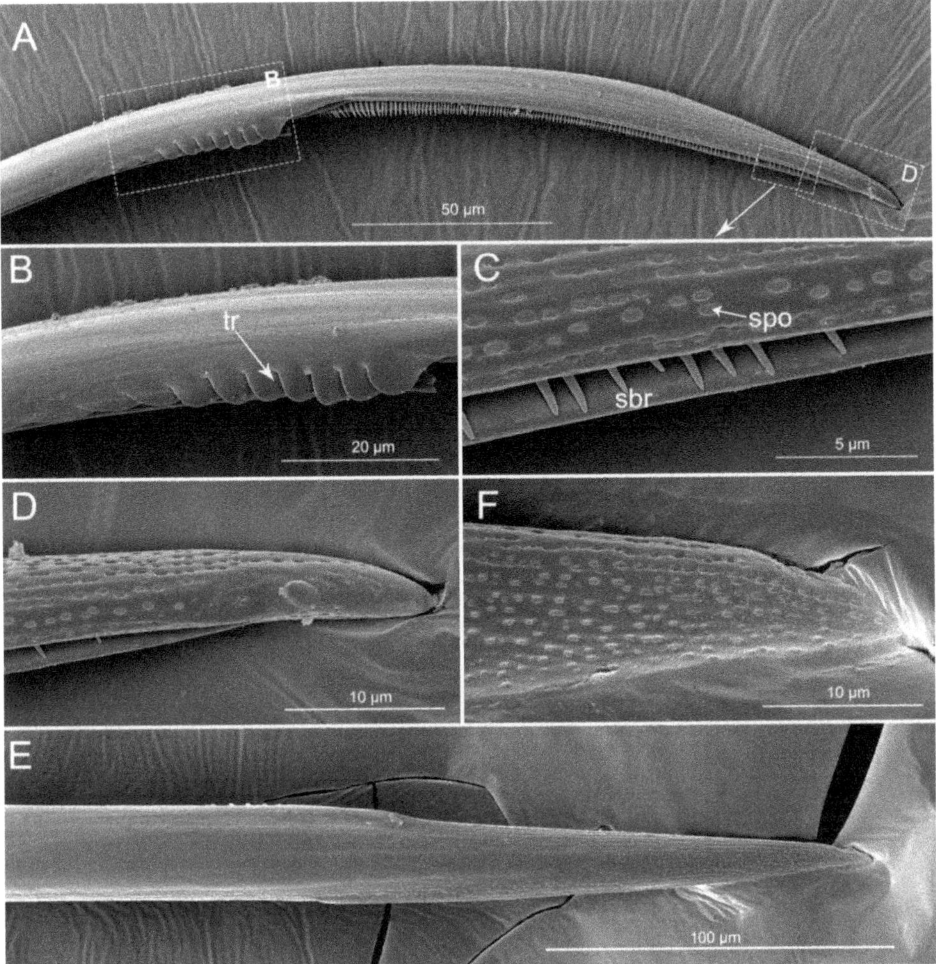

Figure 10. SEM of left maxillary stylets of *Haematoloecha nigrorufa*. (**A**). Lateral view; (**B**). Enlarged view of box in (**A**); (**C**). Enlarged view of box in (**A**); (**D**). Enlarged view of box in (**A**); (**E**). External view; F. Enlarged view of tip of left maxillary stylet; tr, transverse ridges; sbr, short bristles; spo, small pore.

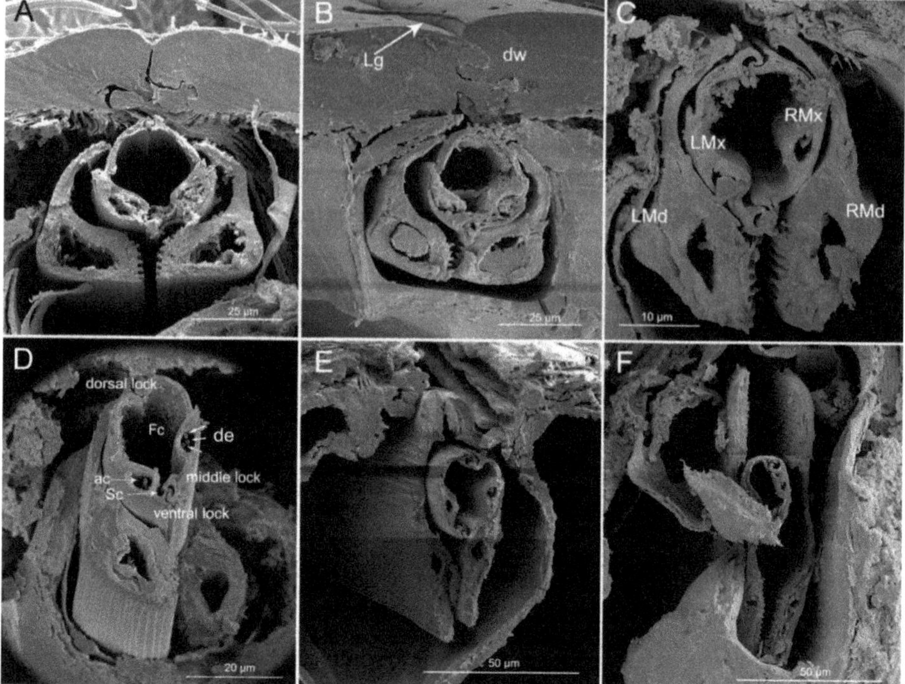

Figure 11. Cross-sections of stylet fascicle and labium of *Haematoloecha nigrorufa*. (**A**). Proximal part of labial segment; (**B**). Median part of segment 2; (**C**). Junction of segments 2 and 3; (**D**). Junction of segments 2 and 3 showing dendrites; (**E**). Median part of segment 3; (**F**). Junction of segments 3 and 4; LMd, left mandibular stylet; LMx, left maxillary stylet; RMd, right mandibular stylet; RMx, right maxillary stylet; Fc, food canal; Sc, salivary canal; de, dendrite; ac, axial canal; Lg, labial groove; dw, dorsal wall.

3.5. The Process of Feeding by Haematoloecha Nigrorufa

H. nigrorufa only feeds on millipedes, according to our observations. Both nymphs and adults of this reduviid exhibit negative phototaxis and are gregarious. The adult feeding process involves several steps, including searching and capturing prey, paralyzing prey, a resting phase, and a feeding phase.

Before insects feed, there is a process of finding and capturing prey. When a hungry predator is looking for prey, if the prey is wandering, the time to identify prey is shorter than when the prey is still, since the movement of the prey attracts the visual and the sensory attention of the predator. When this predator senses the presence of millipedes, it will sway the antennae and turn the body to locate the prey. Once the prey has been located, the predator tracks and slowly moves toward the millipede's head, raises the body, and then captures it as quickly as possible, securing it with the forelegs with stylets oriented toward the substrate. In most cases, the predator captures the prey from behind or from the side.

When the prey is caught, the predator uses the forelegs and midlegs to immobilize the prey and pierces the head-collum intersegmental membrane of the prey with its proboscis. In this process, the predator keeps its proboscis perpendicular to the site of paralysis (Figure 12A). It then presses the tip of the labium onto the paralyzing site and then inserts the stylets (Figure 12B). During this process of anesthetizing, the prey's body contorts acutely, while the predator clings to the millipede's head for three to 10 minutes. Following paralysis, a resting period of five to fifteen minutes occurs. During this period, the predator might crawl around, clean the antennae and rostrum with the forelegs, and drag the immobilized prey. This cleaning of the body is essential in view of the fact that millipedes, which are preys of these insects, discharge a variety of defensive repellant secretions.

Figure 12. Feeding behavior of adult *H. nigrorufa* on millipede. (**A**). Selecting an appropriate paralyzing site; (**B**). Paralyzing millipede; (**C,D**). Feeding on millipede; Lb, labium.

Before the feeding begins, the predator drags the prey to a concealed place. The feeding site of these insects is usually at the intersegmental membrane in the ventral or ventrolateral area. When the predator begins to feed, the beak swings out and the angle between the head and the beak is obtuse (Figure 12C,D). They press the tip of the labium onto the feeding site, insert the stylet, and stay still for about five minutes, and then rest for a while before moving to another part of the body to feed again. The whole feeding process might last from one hour to more than four hours. The labium does not bend and the stylets do not detach from the labium during the entire feeding process. After feeding, the rostrum returns to its resting position along the sternum.

If the predator is disturbed while feeding, it will drag its prey to a safer place and continue feeding. After feeding, it leaves the prey's body behind in order to rest and clean their antennae in a dark hiding place. Most millipede prey are left with just the exoskeleton, with all soft internal tissues having been consumed by the reduviid.

4. Discussion

The overall morphological features of the mouthparts of *H. nigrorufa* are similar to those of other reduviid predators [51,56], including the shape of the labrum, the number of segments of the labium, and the stylet fascicle being composed of two separated mandibular stylets (Md) and two interlocked maxillary stylets (Mx). According to Hwang and Weirauch [57], factors other than microhabitat association may have driven the diversification of Reduviidae; among these are prey specialization and changes in prey capture behavior. In this study of the mouthpart structures in the Reduviidae

(Ectrichodiinae) species *H. nigrorufa*, we focused on detailed morphological features that may be related to their unique adaptation for millipede-feeding.

4.1. The Mouthparts of A Millipede Specialist

The detailed structure of the labrum is rarely discussed in studies of mouthpart morphology. Spooner [48] reported three different shapes of the labrum (broad and flap-like; long, narrow, and triangular, and broad and flap-like with a long epipharyngeal projection) in Heteroptera. Some previous authors showed that the shape of the labrum might be used for characterization of different Heteroptera taxa [46,48–50,52]. In 1969, Štys [50] described several shapes (spiniform, truncate, extremely narrow, and elongate) of the labrum of some Reduviidae, but another unusual condition of the labrum was found in a species of *Ectrychotes* (Ectrichodiinae): the division by a conspicuous transverse membranous zone into two equally well sclerotized parts, called the basilabrum and distilabrum. This labrum structure has been used to characterize the millipede assassin bugs [42,46,52]. In our study we observed that the labrum of *H. nigrorufa* is divided by a transverse membrane into a wide and plicated basilabrum (bl) and an elongated, cone-shaped distilabrum (dl). The non-sclerotized line dividing basilabrum from distilabrum in *Ectrychotes* [50] and other studied taxa (*Nularda nobilitata*, *Ectrichodiella minima*) [42] and *H. nigrorufa* may be, either a novel trait or a remnant of an intermediate stage in the evolution of a long labrum by means of sclerotization of the epipharynx [50]. As suggested by Spooner [48] and Štys [50], a broad, flap-like, simple labrum is probably a primitive feature of Heteroptera. Among reduviids, a subdivided labrum is only characteristic of Ectrichodiinae and Triatominae [42], both unique trophic specialists. Weirauch [42] suggested that a subdivided labrum is of independent origin in Ectrichodiinae and Triatominae, and synapomorphic for both groups. Moreover, the labrum of the studied species shows significantly different composition of sensilla when compared to other reduviid species and this aspect is discussed in the following section on types and functions of sensilla.

The labium of *H. nigrorufa* consists of three visible segments; but, the two subfamilies Hammacerinae and Centrocneminae with four-segmented labia in Reduviidae [42,43,58–60]. In most Reduviidae the first segment is deemed to be either lost or fused to the head capsule [43], which suggested that the four-segmented labium in Hammacerinae is plesiomorphic and homologous to those of non-reduviid Cimicomorpha Weirauch [43]. In most taxa of Heteroptera, the labium is four-segmented, and this feature is used in the classification of the true bug taxa [43,48,60]. The labium of *H. nigrorufa* generally is similar to those of other reduviid species in the number of segments and membrane connections between them. However, another feature in this species is the shape of the last segment. Our SEM observations showed that the segment is short, bent dorsad, and hooklike, differing from other species of Reduviidae, in which the last segment is straight or slightly curved, long or short [61].

We consider the hook-shaped last segment of *H. nigrorufa* to be a special adaptation to feeding on millipedes, although the labium of heteropterans plays an indirect role in predation (maintaining the bundle of stylets, to act as a guide as the stylets are pushed into host/prey tissue) [10,23]. Members of Ectrichodiinae usually approach the millipede's head and paralyze their prey by inserting the stylet at the head-collum intersegmental membrane, according to descriptions of feeding behavior [62–66]. Moreover, if the assassin bug is disturbed, it will drag the prey to a safer place and continue feeding. In both situations, the hook-shaped segment of *H. nigrorufa* seems to be helpful because its curved shape facilitates the perpendicular insertion of the stylets to the membrane between the head and collum, as well as assists with dragging the victim's body. During an attack, the assassin bug clings to the millipede's head (for about four minutes) and the head-collum region is frequently selected to avoid inserting stylets into laterally or dorsally located defensive glands along the trunk during millipede immobilization and consumption [53].

The stylets are the main feeding organs and show great differences among groups with different feeding habits in Heteroptera [12,15,16,19,23,24,35,36,42,45,46,51,56,67–74]. The subfamily of Ectrichodiinae has a strong preference for feeding on millipedes [53,75], but detailed research on the mouthpart structures in this group were previously lacking. We only found general data of morphology of the stylets in previous papers by Cobben [12], Weirauch [42], and Forthman and Weirauch [46] described for some species based on light or SEM microscope studies. The present study reveals several new characters of the mandibular stylets in *H. nigrorufa*, e.g., a greatly expanded apex with abruptly tapered tip and 150 approximately transverse ridges on the external middle side. Moreover, on the inner side of the right and left mandibles are a row of cuticular spikes and longitudinal ridges (lr). These longitudinal ridges are used to lock the two mandibular stylets together below the ventral side of the maxillae; they are more visible in cross section (Figure 11). The maxillary stylets are enclosed in the spaces between the inner longitudinal ridges. A similar spatulate shape of the mandibles is present in other Ectrichodiinae, *Brontosioma discus* Burm [12], and *Nularda nobilitata* [42], but the transverse ridges are usually less numerous (up to 35 or more than 35) and they apparently lack spikes on the inner surface. Similar spatulate mandibles, but lacking the transverse ridges on the outer surface, were observed in some species of Harpactorinae and Sphaeridopinae [12]. The greatest similarity in mandible shape occurs between Ectrichodiinae and Tribelocephalinae [42]; the latter group feeds on ants, termites, and blattids [75]. Both taxa are characterized by rather faint transverse ridges on the outer stylet surface. Based on combined phylogenetic results, Tribelocephalinae were synonymized with Ectrichodiinae [46]. Presumably, transverse ridges help to keep the stylets firmly anchored in prey tissues during the initial attack when the prey may be struggling. Perhaps the larger number of ridges in the studied species is related to the tendency of millipedes to struggle more vigorously after being attacked than the prey of other reduviids with fewer or no ridges.

The left-right asymmetry of the maxillary stylets in *H. nigrorufa* is noticeably similar to that found in most heteropterans, especially predators [12,19]. Previous studies indicate that the right maxillary stylets of predacious bugs usually possess more barb rows [12,15,16,19,42,45,46,51,56,68,69]. Our study is consistent with these previous observations. The characteristic of the right maxilla appear as lamellate structures in the internal ventral side. The lamella are only present in other species of Ectrichodiinae and Tribelocephalinae [42]. Many of the different curved hair-like processes observed in *H. nigrorufa* are also usually present in other true bugs [19,46,68]. Because the hair-like and bristle-like structures are numerous and localized in ventral and dorsal rows (Figure 8A–E), their distribution is similar to those of other heteropterans, except those taxa with reduced hairs, such as Triatominae [42]. The left stylet in the studied species has fewer internal hair-like processes. Frequently the hair system of both maxillae is called a grating-system [12]. Because the food in the suction stylets is semi-fluid substances and particulate matter, which result from the lacerating effects of the spines extending from the maxillae, it must be filtered through the underlying grating structures. In *H. nigrorufa*, bigger spikes and very small edge files are present on the outer side of the left stylet. This mentioned set of hair-like structures is only similar to other Ectrichodiinae. Thus, as suggested by Weirauch [42] and Forthman and Weirauch [46], more extensive comparative studies of the morphology of the maxillary stylets may provide useful taxonomic characters.

Cross sections of the stylets show that the fine-structure and interlocking mechanism of the maxillae of *H. nigrorufa* are similar to those of other reduviid taxa [12,47].

4.2. Labial Sensillar System

Detailed morphological descriptions of Ectrichodiinae (Reduviidae) labial sensilla have never been previously reported. The labium of hemipterans plays an important role in not only receiving the stylet fascicles, but also in the detection of the host by the sensory structures present on the surface [6,10]. The various sensilla, distinguishable by their different external morphological characters perform different functions including chemosensory (gustatory and olfactory), thermo-hygroreceptive, proprioceptive and mechanosensory. In this study, fifteen types of sensilla were observed on the mouthparts of the reduviid predator *H. nigrorufa*.

The most abundant sensilla on the labium in this species are three kinds of sensilla trichodea, which have no pores, a more or less flexible shank, and the base embedded in a flexible socket (St2, St3, and St4), and are considered to be mechanoreceptive. Mechanoreceptors also include sensilla basiconica type Sb1 with a proprioceptive function and sensilla campaniformia (Sca1, Sca 2, and Sca 3). New in this study is the report of sensilla trichodea (St1) with porous walls, which occur in large numbers on the labium and labrum. These sensilla are considered to be chemoreceptive (olfactory). In other species of Ectrichodiina,e the sensilla on the labium were not studied, so it is not yet known whether these sensilla represent a special adaptation that is related to feeding on millipedes. This type of sensilla was also not observed in studies of the labial sensilla in Triatominae and Peiratinae [18,44,76]. The sensitivity and chemical range of insect olfactory systems is remarkable, enabling them to detect and discriminate a wide range of different odor molecules. There is a striking evolutionary convergence towards a conserved organization of signaling pathways in all insect olfactory systems, because the olfactory transduction and neural processing in the peripheral olfactory pathway involve basic mechanisms that are universal across species [77]. Such functioning of the olfactory system does not exclude olfactory sensillae of a different shape than those listed e.g., St1 may also be included in this system. We can only assume that in the studied species (or all species of Ectrichodiinae) only one type of sensilla (St1) on the labium responds to odors of their millipede prey, as suggested by their presence and abundance in these millipede specialists, but absent in related groups of Reduviidae. Unfortunately, the antennal system sensilla in this group of reduviids has not been studied, so data of other types of olfactory sensillae are unknown.

On the labial tip in most heteropterans, there is usually a singule olfactory sensillum basiconicum and/or sensillum placodeum. These sensilla are arranged in a pattern that is highly stereotypical among most heteropterans [15,16,18,19,21,23–32,72,78].

The thermo-hygro receptive sensilla identified as sensilla basiconica (Sb2) and multilobular sensilla (Sm). The flavor sensilla located near the plectrum, which includes three types—Sb4, Sb5, and Sb6—and placoid elongated sensilla (Spe). The described shapes and functions of sensilla, except for sensilla trichoidea (St1) in *H. nigrorufa*, conform to the general model of the labial sensilla present in other reduviids [18,44,76]. However, slight differences may be observed among taxa, especially in the quantity and size of sensilla. The present morphological and functional classifications of the labial sensilla in *H. nigrorufa* are in accordance with the features described for the mentioned types and functions of sensilla reported by many authors [18,24,25,44,54,55,79].

5. Conclusions

This study provides the first detailed fine-structural characterization of the unique mouthparts in *Haematoloecha nigrorufa* (Ectrichodiinae), including the location and distribution of different sensilla types. Judging from the morphology and function, the basal set of types/subtypes of the labial sensilla of *H. nigrorufa* does not strongly differ from other species of reduviids, meaning that a similar pattern of sensilla is visible. However, in particular, we report the presence of more numerous olfactory sensilla trichodea (St1) on the labium and labrum in comparison to the other types of olfactory sensilla in reduviids and other heteropteran taxa. There is also novelty in the special shapes of the labrum, the hook-shaped ultimate segment of the labium, and the large spatulate apex with the many transversal shallow grooves on the external side of the mandibles. These mentioned structures may represent specialized adaptations that are related to the millipede feeding.

Author Contributions: Data curation, W.D.; Funding acquisition, W.D.; Investigation, Y.W., J.Z., W.W.; Methodology, Y.W. and W.D.; Project administration, W.D.; Resources, W.D.; Writing—original draft, Y.W., J.B. and W.D. All authors have read and agreed to the published version of the manuscript.

Funding: This project was supported by the National Natural Science Foundation of China (Nos. 31772514, 31572306, 31272343), the Program of the Ministry of Science and Technology of the People's Republic of China (2015FY210300) and Shaanxi Province Undergraduate Training Program for Innovation and Entrepreneurship (S201910712269).

Acknowledgments: We are grateful to John Richard Schrock (Emporia State University, Emporia, KS, USA) who reviewed the manuscript and offered several valuable suggestions for improvement. We also thank the Life Science Research Core Services of Northwest A&F University for providing scanning electron microscope.

Conflicts of Interest: The authors declare no conflict of interest.

References

1. Snodgrass, R.E. *Principles of Insect Morphology*; McGraw Hill Book Co.: New York, NY, USA, 1935.
2. Chapman, R.F. Communication. In *The Insects: Structure and Function*, 4th ed.; Chapman, R.F., Ed.; Cambridge University Press: Cambridge, UK, 1998; pp. 610–652.
3. Chapman, R.F. Mouthparts. In *Encyclopedia of Insects*; Resh, V.H., Cardé, R.T., Eds.; Academic Press: San Diego, CA, USA, 2003; pp. 750–755.
4. Cummins, K.W. Trophic relations of aquatic insects. *Annu. Rev. Entomol.* **1973**, *1*, 183–206. [CrossRef]
5. Cummins, K.W. Structure and function of stream ecosystems. *Bioscience* **1974**, *2*, 631–641. [CrossRef]
6. Backus, E.A. Sensory systems and behaviours which mediate hemipteran plant-feeding: A taxonomic overview. *J. Insect Physiol.* **1988**, *34*, 151–165. [CrossRef]
7. Krenn, H.W. Feeding mechanisms of adult Lepidoptera: Structure, function, and evolution of the mouthparts. *Annu. Rev. Entomol.* **2010**, *55*, 307–327. [CrossRef]
8. Schneider, D. Insect antennae. *Annu. Rev. Entomol.* **1964**, *9*, 103–122. [CrossRef]
9. Zacharuk, R.Y. Antennae and sensilla. In *Comprehensive Insect Physiology, Biochemistry and Pharmacology*; Kerkut, G.A., Gilbert, L.I., Eds.; Pergamon Press: Oxford, UK, 1985; pp. 1–69.
10. McGavin, G.C. Food and Feeding. In *Bugs of the World*; Blandford: London, UK, 1993.
11. Butt, F.H. Comparative study of mouth parts of representative Hemiptera-Heteroptera. *Ann. Entomol. Soc. Am.* **1943**, *38*, 124. [CrossRef]
12. Cobben, R.H. Evolutionary Trends in Heteroptera. In *Part II Mouthpart-Structures and Feeding Strategies*; Mededelingen Landbouwhogeschool: Wageningen, The Netherlands, 1978; pp. 1–407.
13. Muir, F.; Kershaw, J.C. On the Homologies and Mechanism of the Mouth-Parts of Hemiptera. *A. J. Entomol.* **1911**, *18*, 1–12. [CrossRef]
14. Gullan, P.J.; Cranston, P.S. (Eds.) The Insects: An Outline of Entomology. In *Sensory Systems and Behavior*, 5th ed.; Ch4. Wiley Blackwell: Oxford, UK, 2014.
15. Wang, Y.; Brożek, J.; Dai, W. Sensory armature and stylets of the mouthparts of *Stephanitis nashi* (Hemiptera: Cimicomorpha: Tingidae), their morphology and function. *Micron* **2020**, *132*, 102840. [CrossRef]
16. Wang, Y.; Brożek, J.; Dai, W. Morphological Disparity of the Mouthparts in Polyphagous Species of Largidae (Heteroptera: Pentatomomorpha: Pyrrhocoroidea) Reveals Feeding Specialization. *Insects* **2020**, *11*, 145. [CrossRef]
17. Engel, M.S. A new interpretation of the oldest fossil bee (Hymenoptera: Apidae). *Am. Mus. Novit.* **2000**, *3296*, 1–11. [CrossRef]
18. Brożek, J.; Chłond, D. Morphology, arrangement and classification of sensilla on the apical segment of labium in Peiratinae (Hemiptera: Heteroptera: Reduviidae). *Zootaxa* **2010**, *2476*, 39–52. [CrossRef]
19. Brożek, J. A comparison of external and internal maxilla and mandible morphology of water bugs (Hemiptera: Heteroptera: Nepomorpha). *Zootaxa* **2013**, *3635*, 340–378. [CrossRef] [PubMed]
20. Brożek, J. Phylogenetic signals from Nepomorpha (Insecta: Hemiptera: Heteroptera) mouthparts: Stylets bundle, sense organs, and labial segments. *Sci. World J.* **2014**, *2014*, 237854. [CrossRef] [PubMed]
21. Brożek, J.; Zettel, H. A comparison of the external morphology and functions of labial tip sensilla in semiaquatic bugs (Hemiptera: Heteroptera: Gerromorpha). *Eur. J. Entomol.* **2014**, *111*, 275–297. [CrossRef]
22. Parveen, S.; Ahmad, A.; Brożek, J.; Ramamurthy, V.V. Morphological diversity of the labial sensilla of phytophagous and predatory Pentatomidae (Hemiptera: Heteroptera), with reference to their possible functions. *Zootaxa* **2015**, *4039*, 246–253. [CrossRef]
23. Wang, Y.; Dai, W. Fine structure of mouthparts and feeding performance of *Pyrrhocoris sibiricus* Kuschakevich with remarks on the specialization of sensilla and stylets for seed feeding. *PLoS ONE* **2017**, *12*, e0177209. [CrossRef]
24. Wang, Y.; Li, L.F.; Dai, W. Fine Morphology of the Mouthparts in *Cheilocapsus nigrescens* (Hemiptera: Heteroptera: Miridae) Reflects Adaptation for Phytophagous Habits. *Insects* **2019**, *10*, 17. [CrossRef]
25. Taszakowski, A.; Nowińska, A.; Brożek, J. Morphological study of the labial sensilla in Nabidae (Hemiptera: Heteroptera: Cimicomorpha). *Zoomorphology* **2019**, *138*, 483–492. [CrossRef]

26. Peregrine, D.J. Fine structure of sensilla basiconica on the labium of the cotton stainer, *Dysdercus fasciatus* (Signoret) (Heteroptera: Pyrrhocoridae). *Int. J. Insect Morphol. Embryol.* **1972**, *1*, 241–251. [CrossRef]

27. Schoonhoven, L.M.; Henstra, S. Morphology of some rostrum receptors in *Dysdercus* spp. *Neth. J. Zool.* **1972**, *22*, 343–346. [CrossRef]

28. Avé, D.; Frazier, J.L.; Hatfield, L.D. Contact chemoreception in the tarnished plant bug *Lygus lineolaris*. *Ent. Exp. Appl.* **1978**, *24*, 217–227. [CrossRef]

29. Hatfield, L.D.; Frazier, J.L. Ultrastructure of the labial tip sensilla of the tarnished plant bug, *Lygus lineolaris* (P. De Beauvois) (Hemiptera: Miridae). *Int. J. Insect Morphol. Embryol.* **1980**, *9*, 59–66. [CrossRef]

30. Gaffal, K.P. Terminal sensilla on the labium of *Dysdercus intermedius* Distant (Heteroptera: Pyrrhocoridae). *Int. J. Insect Morphol. Embryol.* **1981**, *10*, 1–6. [CrossRef]

31. Rani, P.U.; Madhavendra, S.S. Morphology and distribution of antennal sense organs and diversity of mouthpart structures in *Odontopus nigricornis* (Stål) and *Nezara viridula* L. (Hemiptera). *Int. J. Insect Morphol. Embryol.* **1995**, *24*, 119–132. [CrossRef]

32. Rani, P.U.; Madhavendra, S.S. External morphology of antennal and rostral sensillae in four hemipteran insects and their possible role in host plant selection. *Int. J. Trop. Insect Sci.* **2005**, *25*, 198–207. [CrossRef]

33. Anderson, W.G.; Hengmoss, T.M.; Baxendale, F.P.; Baird, L.M.; Sarath, G.; Higley, L. Chinch bug (Hemiptera: Blissidae) mouthpart morphology, probing frequencies, and locations on resistant and susceptible germplasm. *J. Econ. Entomol.* **2006**, *99*, 212–221. [CrossRef] [PubMed]

34. Brożek, J. Morphology and arrangement of the labial sensilla of the water bugs. *Bull. Insectol.* **2008**, *61*, 67–168.

35. Depieri, R.A.; Panizzi, A.R. Rostrum length, mandible serration, and food and salivary canals areas of selected species of stink bugs (Heteroptera, Pentatomidae). *Rev. Brasileira De Entomol.* **2010**, *54*, 584–587. [CrossRef]

36. Depieri, R.A.; Siqueia, F.; Panizzi, A.R. Aging and food source effects on mandibular stylets teeth wear of Phytophagous Stink Bug (Heteroptera: Pentatomidae). *Neotrop. Entomol.* **2010**, *39*, 952–956. [CrossRef]

37. Lucini, T.; Panizzi, A.R.; Backus, E.A. Characterization of an EPG Waveform Library for Redbanded Stink Bug, *Piezodorus guildinii* (Hemiptera: Pentatomidae), on Soybean Plants. *Ann. Entomol. Soc. Am.* **2016**, *109*, 198–210. [CrossRef]

38. Backus, E.A.; Serrano, M.S.; Ranger, C.M. Mechanisms of Hopperburn: An overview of insect taxonomy, behavior, and physiology. *Annu. Rev. Entomol.* **2005**, *50*, 125–151. [CrossRef] [PubMed]

39. Backus, E.A. Anatomical and sensory mechanisms of leafhopper and planthopper feeding behavior. In *The leafhoppers and Planthoppers*; Nault, L.R., Rodriguez, J.G., Eds.; John Wiley & Sons: New York, NY, USA, 1985; pp. 163–194.

40. Feir, D.; Beck, S.D. Feeding behavior of the large milkweed bug, *Oncopeltus fasciatus*. *Ann. Entomol. Soc. Am.* **1963**, *56*, 224–229. [CrossRef]

41. Saxena, K.N. Mode of ingestion in a heteropteran insect *Dysdercus koenigii* (F.) (Pyrrhocoridae). *J. Insect Physiol.* **1963**, *9*, 47–71. [CrossRef]

42. Weirauch, C. Cladistic analysis of Reduviidae (Heteroptera: Cimicomorpha) based on morphological characters. *Syst. Entomol.* **2008**, *33*, 229–274. [CrossRef]

43. Weirauch, C. From four- to three-segmented labium in Reduviidae (Hemiptera: Heteroptera). *Acta Entomol. Musei Nationalis Pragae* **2008**, *48*, 331–344.

44. Catalá, S. Sensilla associated with the Rostrum of eight species of Triatominae. *J. Morpho.* **1996**, *228*, 195–201. [CrossRef]

45. Cohen, A.C. Feeding adaptations of some predaceous Hemiptera. *Ann. Entomol. Soc. Am.* **1990**, *83*, 1215–1223. [CrossRef]

46. Forthman, M.; Weirauch, C. Millipede assassins and allies (Heteroptera: Reduviidae: Ectrichodiinae, Tribelocephalinae): Total evidence phylogeny, revised classification and evolution of sexual dimorphism. *Syst. Entomol.* **2017**, *42*, 575–595. [CrossRef]

47. Brożek, J.; Herczek, A. Internal structure of the mouthparts of true bugs (Hemiptera: Heteroptera). *Pol. J. Entomol.* **2004**, *73*, 79–106.

48. Spooner, C.S. The phylogeny of the Hemiptera based on a study of the head capsule. *Illinois Biol. Monogr.* **1938**, *16*, 1–102.

49. Gupta, A.P. A consideration of the systematic position of the Saldidae and the Mesoveliidae (Hemiptera: Heteroptera). *Proc. Ent. Soc. Washington* **1963**, *65*, 31–38.

50. Štys, P. On the morphology of the labrum in Heteroptera. *Acta Ent. Bohemoslov.* **1969**, *66*, 150–158.

51. Kumar, M.; Sahayaraj, K. Gross morphology and histology of head and salivary apparatus of the predatory bug, *Rhynocoris marginatus*. *J. Insect Sci.* **2012**, *12*, 1–12. [CrossRef] [PubMed]

52. Forthman, M.; Weirauch, C. Phylogenetics and biogeography of the endemic Madagascan millipede assassin bugs (Hemiptera: Reduviidae: Ectrichodiinae). *Mol. Phylogenet Evol.* **2016**, *100*, 219–233. [CrossRef]

53. Forthman, M.; Weirauch, C. Toxic associations: A review of the predatory behaviors of millipede assassin bugs (Hemiptera: Reduviidae: Ectrichodiinae). *Eur. J. Entomol.* **2012**, *109*, 147–153. [CrossRef]

54. Altner, H.; Prillinger, L. Ultrastructure of invertebrate chemo-, thermo-, and hygroreceptors and its functional significance. *Int. Rev. Cytol.* **1980**, *67*, 69–139.

55. Frazier, J.L. Nervous system: Sensory system. In *Fundamentals of Insect Physiology*; Blum, M.S., Ed.; John Wiley & Sons: New York, NY, USA, 1985; pp. 287–356.

56. Sahayaraj, K.; Kanna, A.V.; Kumar, S.M. Gross morphology of feeding canal, salivary apparatus and digestive enzymes of salivary gland of *Catamirus brevipennis* (Serville) (Hemiptera: Reduviidae). *J. Entomol. Res. Soc.* **2010**, *12*, 37–50.

57. Hwang, W.S.; Weirauch, C. Evolutionary history of assassin bugs (Insecta: Hemiptera: Reduviidae): Insights from divergence dating and ancestral state reconstruction. *PLoS ONE* **2012**, *7*, e45523. [CrossRef]

58. Schuh, R.T.; Štys, P. Phylogenetic analysis of cimicomorphan family relationships (Heteroptera). *J. N. Y. Entomol. Soc.* **1991**, *99*, 298–350.

59. Schuh, R.T.; Weirauch, C.H.; Wheeler, W.C. Phylogenetic relationships within the Cimicomorpha (Hemiptera: Heteroptera): A total-evidence analysis. *Syst. Entomol.* **2009**, *34*, 15–48. [CrossRef]

60. Weirauch, C.; Bérenger, J.-M.; Berniker, L.; Forero, D.; Forthman, M.; Frankenberg, S.; Freedman, A.; Gordon, E.; Hoey-Chamberlain, R.; Hwang, W.S.; et al. An illustrated identification key to assassin bug subfamilies and tribes (Hemiptera: Reduviidae). *Can. J. Arthropod Identif.* **2014**. [CrossRef]

61. Schuh, R.T.; Slater, J.A. True Bugs of the World (Hemiptera: Heteroptera). In *Classification and Natural History*; Cornell University Press: New York, NY, USA, 1995.

62. Green, W.E. The President's address. *Proc. Entomol. Soc. Lond.* **1925**, *1924*, clxi–ccii.

63. Cachan, P. Etude de la prédation chez les Réduvides de larégion Éthiopienne. 1. La prédationengroupe chez Ectrichodiagigas H-Sch. *Phys. Comp. Oecol.* **1952**, *2*, 378–385.

64. Miller, N.C.E. *The Biology of the Heteroptera*; E. W. Classey: Hampton, VA, USA, 1971; p. 206.

65. Haridass, E.T. Biological and Ethological Studies on some South Indian Reduviids (Hemiptera–Reduviidae). Ph.D. Thesis, University of Madras, Madras, India, 1978.

66. Haridass, E.T.; Ananthakrishnan, T.N. Models for the predatory behavior of some reduviids from southern India (Insecta–Heteroptera–Reduviidae). *Proc. Indian AS Anim. Sci.* **1980**, *89*, 387–402. [CrossRef]

67. Rathore, Y.K. Studies on the mouth-parts and feeding mechanism in *Dysdercus cingulatus* Fabr. (Pyrrhocoridae: Heteroptera). *Indian J. Entomol.* **1961**, *23*, 163–185.

68. Faucheux, M.M. Relations entre l'ultrastructure des stylets manibulaires et maxillaires et la prise de nourriture chez les insects Hemipteres. *CR Acad. Sci. Paris (Ser. D)* **1975**, *281*, 41–44.

69. Boyd, D.W. Digestive enzymes and stylet morphology of *Deraeocoris nigritulus* (Uhler) (Hemiptera: Miridae) reflect adaptations for predatory habits. *Ann. Entomol. Soc. Am.* **2003**, *96*, 667–671. [CrossRef]

70. Swart, C.C.; Felgenhauer, B.E. Structure and function of the mouthparts and salivary gland complex of the giant waterbug, *Belostoma lutarium* (Stål) (Hemiptera: Belostomatidae). *Ann. Entomol. Soc. Am.* **2003**, *96*, 870–882. [CrossRef]

71. Roitberg, B.D.; Gillespie, D.R.; Quiring, D.M.J.; Alma, C.R.; Jenner, W.H.; Perry, J.; Peterson, J.H.; Salomon, M.; Van Laerhoven, S. The cost of being anomnivore: Mandible wear from plant feeding in a true bug. *Naturwissenschaften* **2005**, *92*, 431–434. [CrossRef]

72. Romani, R.; Salerno, G.; Frati, F.; Conti, E.; Isidoro, N.; Bin, F. Oviposition behaviour in *Lygus rugulipennis*: Amorpho-functional study. *Entomol. Exp. Appl.* **2005**, *115*, 17–25. [CrossRef]

73. Bérenger, J.; Pluot-Sigwalt, D. Notes sur *Micrauchenus lineola* (Fabricius 1787), esp cetermitophile et termitophage (Heteroptera: Reduviidae: Harpactorinae, Apiomerini). *Ann. la Soc. Entomol. Fr.* **2009**, *45*, 129–133. [CrossRef]

74. Stubbins, F.L.; Mitchell, P.L.; Turnbull, M.W.; Reay-Jones, F.P.F.; Greene, J.K. Mouthpart morphology and feeding behavior of the invasive kudzu bug, *Megacopta cribraria* (Hemiptera: Plataspidae). *Invertebr. Biol.* **2017**, *136*, 309–320. [CrossRef]

75. Ambrose, D.P. Economic importance of Heteroptera. In *Heteroptera of Economic Importance*; Schaefer, C.W., Panizzi, A.R., Eds.; CRC Press: Boca Raton, FL, USA, 2000; pp. 695–712.

76. Rosa, J.A.D.; Barata, J.M.S.; Cilense, M.; Neto, F.M.B. Head morphology of 1st and 5th instar nymphs of *Triatoma circummaculata* and *Triatoma rubrovaria* (Hemiptera, Reduviidae). *Int. J. Insect Morphol. Embryol.* **1999**, *28*, 363–375. [CrossRef]

77. Hildebrand, J.G.; Shepherd, G.M. Mechanisms of olfactory discrimination: Converging evidence for common principles across phyla. *Annu. Rev. Neurosci.* **1997**, *20*, 595–631. [CrossRef] [PubMed]

78. Baker, G.T.; Chen, X.; Ma, P.W.K. Labial tip sensilla of Blissus leucopterus leucopterus (Hemiptera: Blissidae): Ultrastructure and behavior. *Insect Sci.* **2008**, *15*, 271–275. [CrossRef]

79. Zacharuk, R.Y. Ultrastructure and function of insect chemosensilla. *Annu. Rev. Entomol.* **1980**, *25*, 27–47. [CrossRef]

Article

Resilin Distribution and Sexual Dimorphism in the Midge Antenna and Their Influence on Frequency Sensitivity

Brian D. Saltin [1,2,*], Yoko Matsumura [3], Andrew Reid [1], James F. Windmill [1], Stanislav N. Gorb [3] and Joseph C. Jackson [1]

1 Department of Electronic and Electrical Engineering, Centre for Ultrasonic Engineering, University of Strathclyde, 204 George Street, Glasgow G11 XW, UK; andrew.reid@strath.ac.uk (A.R.); james.windmill@strath.ac.uk (J.F.W.); joseph.jackson@strath.ac.uk (J.C.J.)
2 Department of Biomimetics, Hochschule Bremen—City University of Applied Sciences, Neustadtswall 30, D-28199 Bremen, Germany
3 Department of Functional Morphology and Biomechanics, Zoological Institute of the University of Kiel, Am Botanischen Garten 1–9, D-24118 Kiel, Germany; yoko.matumura.hamupeni@gmail.com (Y.M.); sgorb@zoologie.uni-kiel.de (S.N.G.)
* Correspondence: brian-daniel.saltin@hs-bremen.de

Received: 10 July 2020; Accepted: 3 August 2020; Published: 11 August 2020

Simple Summary: The antennae of insects are multipurpose sensory organs that can detect chemicals, gravity, vibrations, and sound, among others. While such sensors are very specialized and adapted to their specific needs, the way the antenna itself is built has often been considered either uninteresting or unimportant. We used a laser to scan the antenna of the midge *Chironomus riparius*. Insect cuticle, if illuminated with laser light, reflects autofluorescent light, an emission that has long been known to indicate the material properties of the scanned cuticle sample. Rather than a simple beam-like structure of constant material stiffness, we saw bands of hard and soft material, distributed along the length of the antenna. We were able to computer-simulate the effect of this banded structure on the antenna's resonant frequency and showed that it allows the beam to vibrate at different frequencies than would be expected only by its shape. This discovery will help us to better understand these animals' biology and can inspire future biomimetic sensors for detecting sound or vibration.

Abstract: Small-scale bioacoustic sensors, such as antennae in insects, are often considered, biomechanically, to be not much more than the sum of their basic geometric features. Therefore, little is known about the fine structure and material properties of these sensors—even less so about the degree to which the well-known sexual dimorphism of the insect antenna structure affects those properties. By using confocal laser scanning microscopy (CLSM), we determined material composition patterns and estimated distribution of stiffer and softer materials in the antennae of males and females of the non-biting midge *Chironomus riparius*. Using finite element modelling (FEM), we also have evidence that the differences in composition of these antennae can influence their mechanical responses. This study points to the possibility that modulating the elastic and viscoelastic properties along the length of the antennae can affect resonant characteristics beyond those expected of simple mass-on-a-spring systems—in this case, a simple banded structure can change the antennal frequency sensitivity. This constitutes a simple principle that, now demonstrated in another Dipteran group, could be widespread in insects to improve various passive and active sensory performances.

Keywords: *Chironomus riparius*; Diptera; insects; confocal laser scanning microscopy; finite element modelling; antennal hearing; biomechanics; multimodal sensor

1. Introduction

Contrary to mosquitoes, whose bite is not only a nuisance but also a pathway for the transmission of disease, midges receive limited scientific attention. However, midges are numerous in both numbers of individuals and number of species [1] and have been shown to be ecologically important for aquatic and lotic systems [2,3], in terms of biomass and production [4]. The present study on the intricate antennal structure, especially of the male non-biting midge *Chironomus riparius*, aims to reveal some adaptations of these animals' biology.

Chironomus riparius is a non-biting midge that, like many mosquitoes, displays swarming behaviour [5–7]. Since acoustic communications play an essential role in finding mating partners [5,7–9], it is reasonable to expect that there are similarities in the antennal form and hence properties in species of midges and mosquitoes whose mating behaviour includes swarming. Antennae are remarkable sense organs capable of responding to a variety of sense modalities all at once [10,11]. Known functions include senses of smell and gravity, windspeed detection and, in many species, acoustic perception, the latter postulated as long ago as the 19th century [12]. The flagellar nematoceran antenna is built by three elements: most proximally-the scapus, which is partially responsible for orienting the rest of the antenna, followed by the spherical pedicel, housing a Johnston's organ, and most distally the flagellum, which in both sexes appears sub-divided. The number of sensory neurons in the pedicel of mosquitoes has been estimated to be around 16,000 [13]. Most neurons in the Johnston's organ are thought to be involved with acoustic perception, although which ones remains a matter for debate [14].

In males the flagellum is densely covered by fibrillae (also known as setae). These are hair-like structures which are thought to improve sensory performance by increasing the drag of the antenna [8]. Antennae exhibit strong sexual dimorphism, and the female antennae have shorter and fewer fibrillae than the male antennae, which are often referred to as plumose. Despite the known complexity of these auditory systems, mechanical properties of insect sensory organs are often overlooked [15], with just one recent study on the antenna of swarming and non-swarming mosquitoes [16].

To provide another mechanical case study on dipteran antennae, we chose the swarming midge, *Chironomus riparius*. As in the previous study [16], which deployed state-of-the-art confocal laser scanning microscopy (CLSM), the present study presents morphology of the male and female antenna of *C. riparius* through observation of different autofluorescences of varying cuticle configurations. In turn, this study hints at a potential functional influence of the distribution of material composition on resonant tuning of the flagellum. During the last decade, inferring material properties in this way has become an established method [17–23]. CLSM furthermore has the advantage of allowing the imaging of whole structures with no loss of depth resolution, at higher resolutions than conventional light microscopy. In addition to this structural observation, finite element modelling (FEM) of the mechanical behaviour of the antennae with an elasticity distribution in accordance to the observed CLSM data (following the method of [24], see also [16]) shows the potential effect of element position on the mechanical sensitivity. Finally, we discuss the impact of sexual dimorphism of structures and material composition patterns on resonant tuning and its diversity among species in relation to their mating biology.

2. Material and Methods

2.1. Specimen Preparation

Prior to dissection, the animals were anaesthetised with CO_2. Dissection was performed in phosphate buffer solution (PBS) (Carl Roth GmbH & Co KG, Karlsruhe, Germany). The specimens were briefly subjected to small amounts of Triton X-100 (Sigma-Aldrich Chemie GmbH, Steinheim, Germany), to remove air bubbles trapped on the surface by decreasing water surface tension. Triton X-100 then was washed repeatedly with the PBS to fully remove traces of Triton X-100. Microscopical observations were made after transfer of antennae or antennal fragments to glycerine (Carl Roth GmbH & Co. KG, Karlsruhe, Germany).

2.2. Confocal Laser Scanning Microscopy (CLSM)

To analyse local distributing patterns of material compositions within the antenna, we applied CLSM for insect cuticles according to the method established by Michels and Gorb [17]. This technique is successfully used in studies of a wide range of insect exoskeletons [17,23,25,26] including the antennae of mosquitos [16]. The method was applied here as described by Michels and Gorb [17] using a confocal laser scanning microscope, CLSM Zeiss LSM 700 (Carl Zeiss Microscopy GmbH, Jena, Germany). Samples were sequentially exposed to four stable solid-state lasers with wavelengths of 405, 488, 555, and 639 nm, and the excited autofluorescences were filtered with 420–480 nm band-pass and long-pass emission filters transmitting light with wavelengths ≥490, ≥560, and ≥640 nm, respectively. Then, we assigned blue, green, red, and (again) red to the micrographs captured using the filters, respectively, and superimposed them into a final image. To avoid oversaturation, the last two laser lines were combined into one "red" channel, each on 50% intensity. It has to be noted that colours are a product of the colour code applied to the material autofluorescence, and it does not reflect the natural appearance of the antennae. In superimposed images of insect exoskeleton parts, the colour code is as follows: (1) well-sclerotized structures are shown in red, (2) tough flexible cuticular structures are indicated in yellow-green, (3) relatively flexible parts containing a relatively high proportion of resilin appear light-blue and (4) resilin-dominated regions are visualized as deep-blue.

2.3. Finite Element Modelling (FEM)

Finite Element Modelling (FEM) with COMSOL 5.3a (Comsol Inc., Stockholm, Sweden) was conducted to determine the effects of the CLSM results on the mechanical behaviour of the antenna. As with the previous study [16]—where details of the modelling method were already described—the sole purpose of the present simulations is to show that banding and the location of said bands have the potential to influence the beam mechanics. Hence a simplification to a cylinder (10% shell volume) was deemed justifiable to limit computation time, while still encompassing all relevant features of the system.

Similarly, as opposed to stiffness, mass does not tend to be dramatically different between different types of specialised cuticle [27], and therefore it is assumed to be constant in the present simulations. The parts with higher stiffness were simulated with 5 GPa, medium-hard stiffness elements with about 0.5 GPa, and soft material is around 1 MPa (mimicking a typical value for resilin) [27–29].

While the effect of the articulation in the pedicel was not the subject of our study, an approximation was needed and this was achieved by modelling the entire articulations as a round disc at the base of the flagellum, whose flexibility was fixed [16]. For illustration of the basic cylindrical model, please refer to the inset in the FEM simulation figure.

3. Results

In the male *Chironomus riparius*, the pedicel is spherical and exhibits weak autofluorescence in comparison to the rest of the antenna (Figure 1a). The flagellum is composed of 11 units, called flagellomeres. With the exception of the most proximal flagellomere—whose flexible part might be hidden by the pedicel or be part of the articulation—the following ten flagellomeres consist of a basal flexible ring (blue) followed by a sclerotized (red) part, where, except for the 11th flagellomere (Figure 1a *), a circular crest of fibrillae emerges. In the most proximal 11th flagellomere (Figure 1b), fibrillae emerge in an apparently arbitrary pattern. The length of the flagellomeres decreases from the 2nd to 9th, and the lengths are approximately 20–30 µm. The flexible part of the proximal flagellomeres is similar in length to the sclerotized part. In more distal flagellomeres approaching the 10th flagellomere, the flexible part decreases in length to about half of the length of the sclerotized part. The sclerotized part, which is approximately similar in length, gradually loses the dominance of red autofluorescence. From the 5th or 6th flagellomeres onwards, their autofluorescence becomes entirely green (i.e., tough and flexible). The whole structure tapers continuously from the base to the 10th flagellomere—the diameter of the flagellomeres decreases from around 65 μm to 40 μm and continues to taper towards a pointed tip. After the 10th flexible ring (showing strongly blue autofluorescence), the antenna shows less intense autofluorescence until the tip. The fibrillae continuously become shorter along the flagellum up to the very short and irregular fibrillae at the tip. Along the flagellum, none of the fibrillae exhibits any strong autofluorescence (Figure 1a).

As in the other species previously investigated [16], all prongs are of the same diameter and show homogeneous green autofluorescence. The red-orange autofluorescence of the rim of the pedicel, already visible from the outside (Figure 1a), is also visible from the inside (Figure 1c). This indicates that the rim is relatively well sclerotized. The ridge, where the prongs attach, is deep-blue autofluorescent and possibly resilin-enriched, which is not encountered in any other species studied, while the prongs between attachment and flagellum appear to be of stiffer material indicated by reddish autofluorescence (Figure 1c).

In the female *C. riparius*, the pedicel is slightly rectangular in shape and exhibits comparatively strong green fluorescence (Figure 1d). The flagellum is composed of five flagellomeres, which are cylindrical but not constant in diameter within a flagellomere. All flagellomere have 6–8 separate long fibrillae emerging in a crest. There are rings of blue fluorescence (Figure 1d), which are likely resilin-enriched for flexibility. There is also another crest of shorter fibrillae present on each flagellomere. The long fibrillae emerge in one crest at the widest part of each flagellomere as it broadens, before the flagellomere tapers again. The bottom of each flagellomere, with diameters of 40 to 50 μm, tapers to about half this width (Figure 1c,d).

In the first and second flagellomere of *C. riparius*, fibrillae sockets of the fibrillae crest are apparent and are distinctly more orange/red than its remainder. To the right in Figure 1d, the optical section shows the articulation of the flagellum: no further internal details are visible. In comparison to the male and to the other species, the articulation is more flattened than domed (Figure 1d). An optical section (Figure 1e) of the pedicel shows a rather flexible soft articulation with a central blue area (Figure 1e). The pedicel as such is more fluorescent than the flagellum.

The male pedicel (Figure 1a,c), described in detail above, has a more detailed substructure than the female pedicel (Figure 1d,e). The female pedicel, described in detail above (Figure 1d,e), is in general more angular and less spherical, and the articulation of the flagellum is rather flat. There are clear differences in the flagellum's subdivision in to flagellomeres and material distribution along the flagellum between sexes. Female antennae have five flagellomere, in contrast to the eleven flagellomere of the males. In both sexes, the hard parts of the flagellomeres are separated by blue-fluorescent joints. Furthermore, the female antenna is less covered by fibrillae (Figure 1a,d), which in both sexes similarly show relatively weak autofluorescences. The female does not exhibit the characteristic short intervals among green, red-orange, and light blue bands observed in the first ten antennal flagellomeres in males. Instead, the fewer flagellomeres are more evenly spaced out along the whole length of the female antennae.

Based on the observed flagellomere distribution and material distributions, FEM simulations of the mechanical response of a beam structure with and without the revealed substructure were performed (Figure 2). For the shorter female structure (green lines in Figure 2), no effect of the banded structure on the resonant frequency is seen. For the male antenna (blue lines in Figure 2) a small downwards shift of 4 Hz between the uniform structure (continuous light-blue line) and the more realistic substructured beam (dashed, dark blue line) can be observed. In the lower right corner, a more detailed 1 Hz step simulation of the frequency range around the strongest response 445–475 Hz in males is shown.

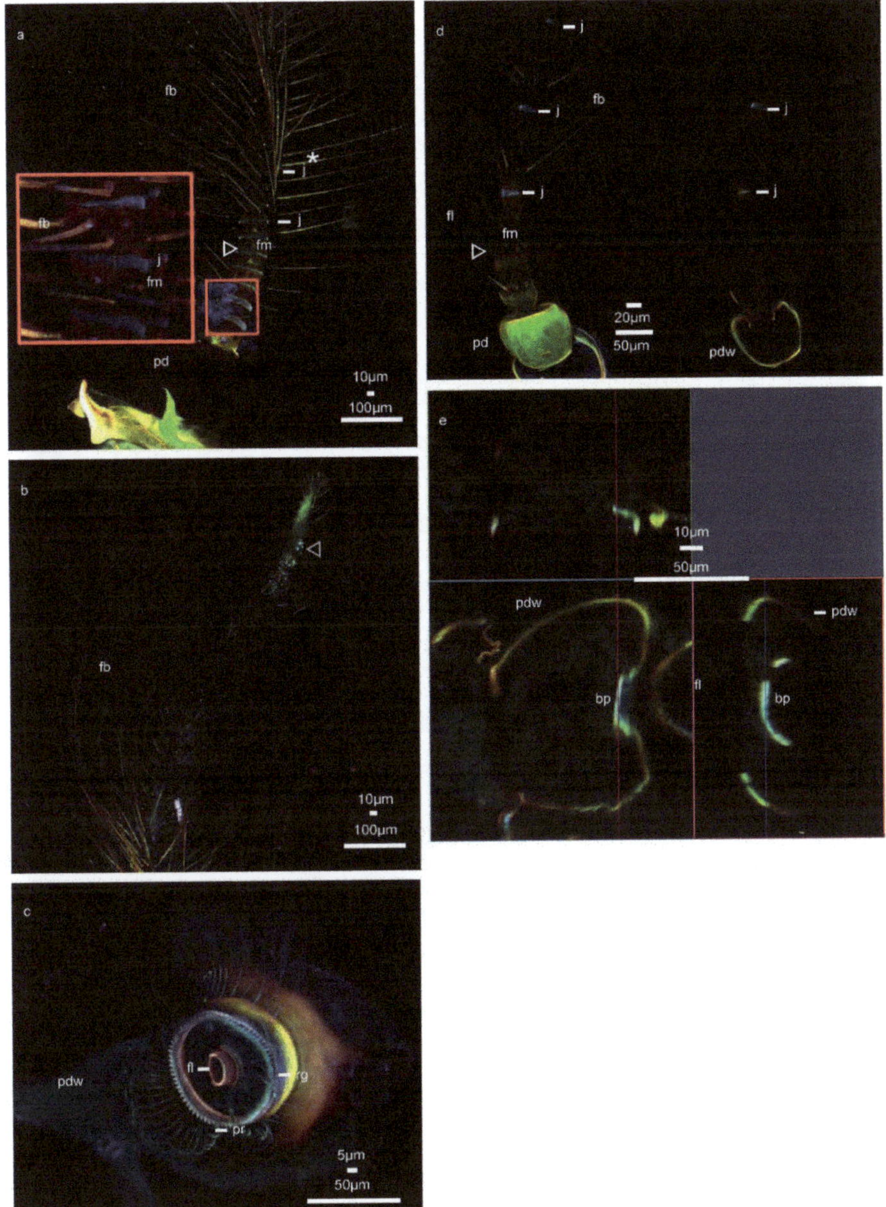

Figure 1. CLSM observations of the antenna of *C. riparius*. The colour code runs from blue colours for comparatively soft structures to increasingly stiff structures in red. (**a**) Maximum intensity projection of the male *C. riparius* antenna, pedicel and basal antenna part with magnified inset. (**b**) Maximum intensity projection of the male *C. riparius* antenna, tip region. (**c**) Maximum intensity projection of the male *C. riparius* pedicel, seen from inside. (**d**) Left: Maximum intensity projection of the female *C. riparius* antenna, right: cross-section. (**e**) Optical cross-section of the female *C. riparius* pedicel. Abbreviations: bp: basal plate, fl: flagellum, fm: flagellomere, fb: fibrillae, j: joint, pd: pedicel, pdw: pedicel wall, pr: prongs, rg: ridge.

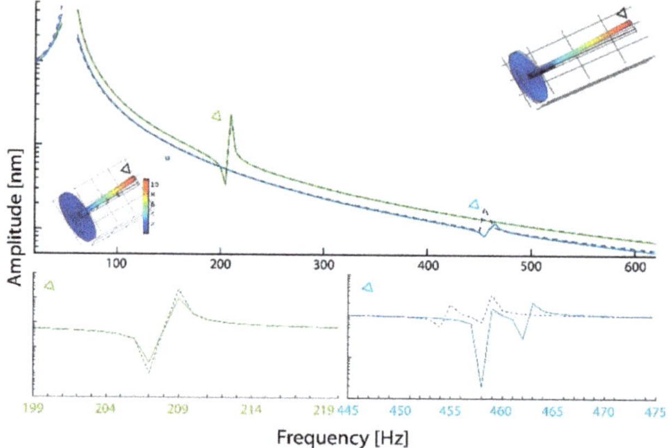

Figure 2. Simulation results for models of uniform and structured female and male *C. riparius* antenna in 5 Hz steps between 20 and 620 Hz. The figure includes for illustrative purposes the simulated female and male model (left and right, respectively). In each model the point (node), whose displacement is shown in the figure, is marked with a black triangle. The displacement is codified as gradient from low (dark blue) to high (red). Indicated in green (female) and blue (male), triangles point to the frequency of the strongest mechanical response. The figures underneath show the zoomed-in response for the female (left, green) and the male (right, blue). In each panel the comparison between a uniform beam of the sex-specific dimension depicted as solid line and the more natural situation of a substructured flagellum-beam depicted as dashed line, also in the sex-specific dimension. The used distribution of substructure is deduced from the antenna CLSM images Figure 1a,b (male) and Figure 1d (female). Simulating an impinging sound field, load was applied perpendicular to the beam axis in the +X direction on all but the lowest element.

4. Discussion

Our results show that the well-known sexual dimorphism of dipteran antennae goes further than morphological structure alone, and in midge antennae also includes differences in material elasticity. Like in mosquitoes [16], material composition of the antennae is not homogeneous along the flagellum, but instead comprises hard and soft elements. Taken together with the structural complexity of the antenna in mosquitoes [16] and stick insects [30], it is becoming more evident that the structure and especially material and composition of insect antennae is much more complex as previously thought. Despite the statement that material properties of insect sensors are largely overlooked made as early as 2009 by Sane and McHenry [15], only limited research has been conducted to amend this lack of understanding. A lot of questions remain open and there is much potential for future research given the vast diversity of insect antennae not yet sampled. To our knowledge, none of the hitherto investigated species here and in our previous study [16] closely resemble each other regarding material distribution irrespective of their mating ecology. This indicates that further research will be necessary to better understand the various factors influencing antenna morphology. Given the different sensory functions of insect antennae that include, but are by far not limited to, olfaction, tactile sensation, and hearing, it is clear that the structure balances various trade-offs and functional constraints. One example of intricate structures of unknown function is the rapid sequence of flexible and sclerotized material at the base of the male flagellum of *C. riparius*.

Pedicels have a large variation of autofluorescence intensity and are largest in *C. riparius* females. In male *C. riparius*, a hard area on the distal ridge, where the flagellum emerges from the pedicel, is most prominently visible. The two important messages regarding the prongs are as follows. First, the prongs are neither particularly flexible nor stiff and are all amongst each other consistent in their autofluorescence within an individual animal. Secondly, judging from our CLSM images, they seem rather similar in dimensions. The uniformity of the prongs in their stiffness and dimension underpins previous assumptions by Avitabile et al. [31] that the prongs act more or less as rigid-body extensions of the flagellum. The flagellum, however, is by virtue of stiffness variation, shown by our study, potentially acts in a more complex manner than simply rigid beam of uniform stiffness. Similar to our study on mosquito antennae [16], we confirmed here the presence of variation and increased small-scale complexity of the dipteran antenna.

While the degree of effect remains under dispute, the direct fitness improvement of traits involved in sexual selection is not [32–34]. An impact of these differences on mating behaviour seems likely given the combination of the following three points: (1) certain mosquitoes (7–9) and at least some midges [5] respond to acoustic stimuli; (2) their antennae clearly show a well-known structural sexual dimorphism (e.g., [35]), and as demonstrated here also a dimorphism in material composition; (3) considerations by Loudon [36] heavily imply the importance of getting the flexural stiffness of antenna right for any given insect. This means that while the function of the different banded structure between species [16] and sexes reported remains unclear for now, they will be meaningful for the behaviour and biology of those animals.

Possible reasons for these antennal observations are that a different stiffness will inadvertently correspond to a different resonant tuning for acoustic perception, or for reasons of static integrity of the antenna, or perhaps another behavioural or ecological aspect of these animals' biology. Compared to results in mosquitoes [16], the effect might be smaller in male Chironomidae or different in principle—both hypotheses require further investigation. Whatever the ultimate reasons for the observed specialisation are, it is fairly clear that different specialisations of males and females might require strong tuning of their acoustic sensors (antennae), which is not understood yet, but this study shows further evidence for the presence of such a specialisation. A limitation of both these studies is the lack of direct correlation of CLSM-based autofluorescence analysis with mechanical measurements, which should be tackled in follow-up investigations.

5. Conclusions

We have demonstrated that the sexual dimorphism in the antenna of *Chironomus riparius* pertains beyond geometry to material composition. The antennae of both sexes balance a variety of functions. Hence it is difficult to decide—without further research—how much of the newly found complexity actually is adaptive to a given sensory function. While effects on resonant tuning in male midges are small compared to the hundreds of Hz shifts observed in mosquitoes [16], variation in stiffness can alter the antenna's vibrational characteristics in different species.

This result and other studies on the mechanics of antennae [16,30] as well as other appendages [26], underlines the necessity of a more holistic and realistic future approach not only but especially for modelling. That includes the hitherto unknown material complexity in these structures.

Future studies of insect antenna could include investigations on other species or be combined with direct mechanical measurements, such as bending and indentation tests, which would provide better understanding of their structure-function relationships. Such outcomes will improve the quality of simulation results, as we clearly see how the mechanical responses can deviate due to structural and material complexities so far observed. The importance of knowledge about material properties of insect cuticle for understanding functional mechanisms of different organs is huge [21,26,30,37–41] e.g., for robotics [40,41], and can be extended to the sensory structures [16,41–45]. This is not only a matter of academic interest but could also feature in the improvement of biomimetic sensory systems with wide applications. Rather than trying to find materials with a given Young's modulus to satisfy a design constraint, stiffness can be altered through careful design of banding with standard materials.

Author Contributions: B.D.S. and Y.M. carried out preparations and CLSM imaging. A.R. and B.D.S. conducted FEM simulations. B.D.S., J.F.W., S.N.G. and J.C.J. designed the study. B.D.S., A.R., S.N.G., Y.M., and J.C.J. wrote the manuscript. All authors have read and agreed to the published version of the manuscript.

Funding: This work was supported in part by the European Research Council under the European Union's Seventh Framework Programme FP/2007-2013/ERC under Grant Agreement n. 615030 to J.F.C.W. This work was partially supported by the European Research Council under the European Union's Seventh Framework Programme FP/2007-2013/ERC under grant agreement no. 615030 to J.F.C.W. by the EPSRC (J.C.J., EP/H02848X/1) and by the German Research Foundation (Y.M., DFG grant no. MA 7400/1-1). B.D.S. is funded by the HSB Research Fellowship. The APC were funded by UKRI.

Acknowledgments: We thank Jan Michels (University of Kiel) for theoretical and practical CLSM training, as well as members of staff of the Centre for Ultrasonic Engineering at the University of Strathclyde for their support, especially Jeremy Gibson. Thanks also go to Carla Lorenz and Heinz-R. Köhler and the Animal Physiological Ecology group (University of Tübingen) for provision of *Chironomus* eggs to start a local culture.

Conflicts of Interest: The authors declare no conflict of interest.

References

1. *The Chironomidae: Biology and Ecology of Non-Biting Midges*, 1st ed.; Armitage, P.D.; Pinder, L.C.; Cranston, P.S., Eds.; Springer Science & Business Media: Berlin/Heidelberg, Germany, 2012.

2. Dévai, G. Ecological background and importance of the change of chironomid fauna (Diptera: Chironomidae) in shallow Lake Balaton. In *Developments in Hydrobiology*; Dumont, H.J., Ed.; Springer: Heidelberg, Germany, 1990; Volume 53, pp. 189–198.

3. Berg, M.B.; Hellenthal, R.A. The role of Chironomidae in energy flow of a lotic ecosystem. *Neth. J. Aquat. Ecol.* **1992**, *26*, 471–476. [CrossRef]

4. Huryn, A.D.; Wallace, J.B. A method for obtaining in situ growth rates of larval Chironomidae (Diptera) and its application to studies of secondary production 1. *Limnol. Oceanogr.* **1986**, *31*, 216–221. [CrossRef]

5. Downes, J.A. The swarming and mating flight of diptera. *Annu. Rev. Entomol.* **1969**, *14*, 271–298. [CrossRef]

6. Caspary, V.G.; Downe, A.E.R. Swarming and mating of Chironomus riparius (Diptera: Chironomidae). *Can. Entomol.* **1971**, *103*, 444–448. [CrossRef]

7. Belton, P. An analysis of direction finding in male mosquitoes. In *Experimental Analysis of Insect Behavior*; Springer: Berlin/Heidelberg, Germany, 1974; pp. 139–148.

8. Göpfert, M.C.; Briegel, H.; Robert, D. Mosquito hearing: Sound-induced antennal vibrations in male and female Aedes aegypti. *J. Exp. Biol.* **1990**, *202*, 2727–2738.

9. Cator, L.J.; Arthur, B.J.; Harrington, L.C.; Hoy, R.R. Harmonic convergence in the love songs of the dengue vector mosquito. *Science* **2009**, *323*, 1077–1079. [CrossRef]

10. Kamikouchi, A.; Inagaki, H.K.; Effertz, T.; Hendrich, O.; Fiala, A.; Göpfert, M.C.; Ito, K. The neural basis of Drosophila gravity-sensing and hearing. *Nature* **2009**, *458*, 165–171. [CrossRef]

11. Matsuo, E.; Kamikouchi, A. Neuronal encoding of sound, gravity, and wind in the fruit fly. *J. Comp. Physiol. A* **2013**, *199*, 253–262. [CrossRef]

12. Johnston, C. Original communications: Auditory apparatus of the Culex mosquito. *Q. J. Microsc. Sci.* **1855**, *1*, 97–102.

13. Boo, K.S.; Richards, A.G. Fine structure of the scolopidia in the Johnston's organ of male Aedes aegypti (L.) (Diptera: Culicidae). *Int. J. Insect Morphol. Embryol.* **1975**, *4*, 549–566. [CrossRef]

14. Hart, M.; Belton, P.; Kuhn, R. The Risler Manuscript. *Eur. Mosq. Bull.* **2011**, *29*, 103–113.

15. Sane, S.P.; McHenry, M.J. The biomechanics of sensory organs. *Integr. Comp. Biol.* **2009**, *49*, i8–i23. [CrossRef]

16. Saltin, B.D.; Matsumura, Y.; Reid, A.; Windmill, J.F.; Gorb, S.N.; Jackson, J.C. Material stiffness variation in mosquito antennae. *J. R. Soc. Interface* **2009**, *16*, 20190049. [CrossRef] [PubMed]

17. Michels, J.; Gorb, S.N. Detailed three-dimensional visualization of resilin in the exoskeleton of arthropods using confocal laser scanning microscopy. *J. Microsc.* **2012**, *245*, 1–16. [CrossRef]

18. Peisker, H.; Michels, J.; Gorb, S.N. Evidence for a material gradient in the adhesive tarsal setae of the ladybird beetle Coccinella septempunctata. *Nat. Commun.* **2013**, *4*, 1661. [CrossRef]

19. Willkommen, J.; Michels, J.; Gorb, S.N. Functional morphology of the male caudal appendages of the Damselfly Ischnura elegans (Zygoptera: Coenagrionidae). *Arthropod Struct. Dev.* **2015**, *44*, 289–300. [CrossRef]

20. Filippov, A.E.; Matsumura, Y.; Kovalev, A.E.; Gorb, S.N. Stiffness gradient of the beetle penis facilitates propulsion in the spiraled female spermathecal duct. *Sci. Rep.* **2016**, *6*, 27608. [CrossRef]

21. Michels, J.; Appel, E.; Gorb, S.N. Functional diversity of resilin in Arthropoda. *Beilstein J. Nanotechnol.* **2016**, *7*, 1241–1259. [CrossRef] [PubMed]
22. Schmitt, M.; Büscher, T.H.; Gorb, S.N.; Rajabi, H. How does a slender tibia resist buckling? Effect of material, structural and geometric characteristics on buckling behaviour of the hindleg tibia in stick insect postembryonic development. *J. Exp. Biol.* **2018**, *221*, 1–4. [CrossRef]
23. Bergmann, P.; Richter, S.; Glöckner, N.; Betz, O. Morphology of hindwing veins in the shield bug Graphosoma italicum (Heteroptera: Pentatomidae). *Arthropod Struct. Dev.* **2018**, *47*, 375–390. [CrossRef]
24. Eshghi, S.H.; Jafarpour, M.; Darvizeh, A.; Gorb, S.N.; Rajabi, H. A simple, high-resolution, non-destructive method for determining the spatial gradient of the elastic modulus of insect cuticle. *J. R. Soc. Interface* **2018**, *15*, 20180312. [CrossRef] [PubMed]
25. Neff, D.; Frazier, S.F.; Quimby, L.; Wang, R.-T.; Zill, S. Identification of resilin in the leg of cockroach, Periplaneta americana: Confirmation by a simple method using pH dependence of UV fluorescence. *Arthropod Struct. Dev.* **2000**, *29*, 75–83. [CrossRef]
26. Matsumura, Y.; Kovalev, A.E.; Gorb, S.N. Penetration mechanics of a beetle intromittent organ with bending stiffness gradient and a soft tip. *Sci. Adv.* **2017**, *3*, 5469–5477. [CrossRef] [PubMed]
27. Vincent, J.F.; Wegst, U.G. Design and mechanical properties of insect cuticle. *Arthropod Struct. Dev.* **2004**, *33*, 187–199. [CrossRef]
28. Wegst, U.G.K.; Ashby, M.F. The mechanical efficiency of natural materials. *Philos. Mag.* **2004**, *84*, 2167–2181. [CrossRef]
29. Kovalev, A.; Filippov, A.; Gorb, S.N. Slow viscoelastic response of resilin. *J. Comp. Physiol. A* **2018**, *204*, 409–417. [CrossRef]
30. Rajabi, H.; Shafiei, A.; Darvizeh, A.; Gorb, S.N.; Duerr, V.; Dirks, J.-H. Both stiff and compliant: Morphological and biomechanical adaptations of stick insect antennae for tactile exploration. *J. R. Soc. Interface* **2018**, *15*, 20180246. [CrossRef]
31. Avitabile, D.; Homer, M.; Champneys, A.R.; Jackson, J.C.; Robert, D. Mathematical modelling of the active hearing process in mosquitoes. *J. R. Soc. Interface* **2010**, *7*, 105–122. [CrossRef]
32. Darwin, C. *The Origin of Species*; The Temple Press: Letchworth, UK, 1928; reprinted 1934.
33. West-Eberhard, M.J. Sexual selection, competitive communication and species specific signals in insects. In *Insect Communication, Proceedings of the 12th Symposium of the Royal Entomological Society of London, London, UK, 7–9 September 1983*; Academic Press: New York, NY, USA, 1984.
34. Laland, K.; Uller, T.; Feldman, M.; Sterelny, K.; Müller, G.B.; Moczek, A.; Jablonka, E.; Odling-Smee, J.; Wray, G.A.; Hoekstra, H.E.; et al. Does evolutionary theory need a rethink? *Nature News* **2014**, *514*, 161. [CrossRef]
35. Gerhardt, R.R.; Hribar, L.J. Flies (Diptera). In *Medical and Veterinary Entomology*, 3rd ed.; Academic Press: Cambridge, MA, USA, 2019; pp. 171–190.
36. Loudon, C. Flexural stiffness of insect antennae. *Am. Entomol.* **2005**, *51*, 48–49. [CrossRef]
37. Dickinson, M.H.; Lehmann, F.-O.; Sanjay, P.; Sane, S.P. Wing rotation and the aerodynamic basis of insect flight. *Science* **1999**, *284*, 1954–1960. [CrossRef] [PubMed]
38. Pennisi, E. Bendy bugs inspire roboticists. *Science* **2016**, *351*, 647. [CrossRef] [PubMed]
39. White, Z.W.; Vernerey, F.J. Armours for soft bodies: How far can bioinspiration take us? *Bioinspiration Biomim.* **2018**, *13*, 041004. [CrossRef] [PubMed]
40. Kaneko, M.; Ueno, N.; Tsuji, T. Active Antenna-basic considerations on the working principle. In Proceedings of the IEEE/RSJ International Conference on Intelligent Robots and Systems (IROS'94), Munich, Germany, 12–16 September 1994; Volume 3, pp. 1744–1750. [CrossRef]
41. Cowan, N.J.; Ma, E.J.; Cutkosky, M.; Full, R.J. A biologically inspired passive antenna for steering control of a running robot. In *Robotics Research. The Eleventh International Symposium*; Springer: Berlin/Heidelberg, Germany, 2005.
42. Barth, F.G.; Németh, S.S.; Friedrich, O.C. Arthropod touch reception: Structure and mechanics of the basal part of a spider tactile hair. *J. Comp. Physiol. A* **2004**, *190*, 523–530. [CrossRef] [PubMed]
43. Schaber, C.F.; Gorb, S.N.; Barth, F.G. Force transformation in spider strain sensors: White light interferometry. *J. R. Soc. Interface* **2012**, *9*, 1254–1264. [CrossRef] [PubMed]

44. Schaber, C.F.; Barth, F.G. Spider joint hair sensilla: Adaptation to proprioreceptive stimulation. *J. Comp. Physiol. A* **2015**, *201*, 235–248. [CrossRef]
45. Sutton, G.P.; Clarke, D.; Morley, E.L.; Robert, D. Mechanosensory hairs in bumblebees (Bombus terrestris) detect weak electric fields. *Proc. Natl. Acad. Sci. USA* **2016**, *113*, 7261–7265. [CrossRef]

 *insects*

Article

Adhesion Performance in the Eggs of the Philippine Leaf Insect *Phyllium philippinicum* (Phasmatodea: Phylliidae)

Thies H. Büscher *, **Elise Quigley and Stanislav N. Gorb**

Department of Functional Morphology and Biomechanics, Institute of Zoology, Kiel University, Am Botanischen Garten 9, 24118 Kiel, Germany; eliseq2@scarletmail.rutgers.edu (E.Q.); sgorb@zoologie.uni-kiel.de (S.N.G.)
* Correspondence: tbuescher@zoologie.uni-kiel.de

Received: 12 June 2020; Accepted: 25 June 2020; Published: 28 June 2020

Abstract: Leaf insects (Phasmatodea: Phylliidae) exhibit perfect crypsis imitating leaves. Although the special appearance of the eggs of the species *Phyllium philippinicum*, which imitate plant seeds, has received attention in different taxonomic studies, the attachment capability of the eggs remains rather anecdotical. We herein elucidate the specialized attachment mechanism of the eggs of this species and provide the first experimental approach to systematically characterize the functional properties of their adhesion by using different microscopy techniques and attachment force measurements on substrates with differing degrees of roughness and surface chemistry, as well as repetitive attachment/detachment cycles while under the influence of water contact. We found that a combination of folded exochorionic structures (pinnae) and a film of adhesive secretion contribute to attachment, which both respond to water. Adhesion is initiated by the glue, which becomes fluid through hydration, enabling adaption to the surface profile. Hierarchically structured pinnae support the spreading of the glue and reinforcement of the film. This combination aids the egg's surface in adapting to the surface roughness, yet the attachment strength is additionally influenced by the egg's surface chemistry, favoring hydrophilic substrates. Repetitive detachment and water-mediated adhesion can optimize the location of the egg to ensure suitable environmental conditions for embryonic development. Furthermore, this repeatable and water-controlled adhesion mechanism can stimulate further research for biomimeticists, ecologists and conservationalists.

Keywords: attachment; glue; oviposition; biomechanics; walking leaf; morphology; plant surface interactions; insect–plant relations; egg dispersal

1. Introduction

Stick insects (Phasmatodea) are rather large terrestrial herbivores and well known for their remarkable camouflage [1,2]. This masquerade, imitating parts of their environment, is particularly striking in the lineage Phylliidae (leaf insects). Consequently, these insects are commonly called "walking leaves" [3–5]. Leaf insects extraordinarily imitate the leaves of plants and visually merge with their environment. The first fossil records of Phylliidae date back 47 mya with *Eophyllium messelensis* Wedmann, Bradler and Rust 2006 as the oldest known representative of this lineage [6]. Visual camouflage in stick insects had already evolved during the Cretaceous period (approximately 125 mya), to avoid predators at a time when gymnosperm plants represented the majority of plant diversity [6–8]. During the emergence of angiosperms and their major radiation [9,10], stick insects evolved in a similar rapid fashion, possibly as a response to the burgeoning diversity of plants [8,11–14]. Camouflage is not only used by the insects to deceive predators, but also exhibited by the eggs in their resemblance to plant seeds [1,13,15]. Beyond visual aspects, ranging from the imitation of twigs, bark, moss and other environmental elements, along with the convergent evolution of leaf mimicry in Phyliidae and several other groups of stick insects [16], other characteristics diversified as well. The attachment systems of phasmids, for example, adapted to the abundance of different plant surfaces [5,17–21]. Females also made use of a remarkably broad range of oviposition techniques, which differ between species depending on their ecological niche [2,5]. As a result, the egg morphology reflects an oviposition technique and ecological niche [15,22,23]. Some species simply drop their eggs passively while others catapult them actively. Passively dropping the eggs is considered an extension of phasmids' notorious masquerade crypsis [2] and probably an ancestral technique [2,13,24]. Another widespread principle is fixation of one egg or groups of eggs to specific spots, e.g., their host plants [13]. While some species mechanically drill their eggs into the soil, into crevices or even leaves and bark, other species secrete a glue during oviposition to permanently fix the eggs to the substrate [2,5,13,20]. The latter is either used to attach single eggs or batches to a certain place, and in one striking case, the egg batches are deposited in the form of an ootheca with a protective case [13]. The different strategies for egg deposition are a result of the low spatial distribution and extensive radiation of phasmids, which presumably led to the co-evolution with angiosperms.

Interestingly, not only has the outer appearance of phasmids been shaped by their co-existence with plants, but also the eggs of phasmids mimic the appearance of seeds and even copy functional principles of seeds. Phasmids are not only the sole insect lineage with species-specific egg appearances [2], but also the only lineage with eggs adapted to different oviposition techniques. Some taxa, in which the eggs are dropped passively, produce eggs which bear a capitulum. This extension of the egg's operculum is not only a signal adaptation for zoochory by ants, but also a result of co-evolution with plant seeds and ants [25,26]. In both capitulate phasmid eggs and elaiosome-bearing seeds, such a lipid-rich extension mimics ant-specific signaling and convinces ants to carry the egg or seed and thereby mediate dispersal [27–29]. Besides ant-mediated zoochory, the eggs of several species of phasmids follow the same principles that plant seeds deploy for dispersion, or aggregation respectively. Many plant seeds disperse via endozoochory, especially via birds [30,31]. Although an initial study has shown that phasmid eggs (directly fed to birds) of a few species do not survive the digestion by quills and ducks [32], a subsequent study found the eggs of several other phasmid species remain viable inside a gravid female phasmid that has been consumed by a bird [33]. Other phasmid eggs, especially *Megacrania* species, are experimentally shown to float in sea water, and disperse via the ocean [34–37], like the seeds of *Cycas* spp. (Cycadaceae) or screw pines do [38,39].

Phyllium philippinicum Hennemann, Conle, Gottardo and Bresseel, 2009 (Phylliidae) is a species of leaf insect commonly bred in labs and private cultures (Figure 1A). However, most of the literature on the species revolves around taxonomic and phylogenetic classification and is mainly based on adult morphology [3,40–42]. Leaf insects in general are reported to drop or catapult their egg for deposition from the canopy tops of their host tree [2,13]. Basically, the eggs of this species, as well as those of closely related species, employ a more specialized mechanism for host plant association than previously reported. The specialized exochorionic morphology of leaf insect eggs is predominantly accounted in descriptive morphological studies and taxonomic descriptions [4,15,23,42] and, hence, functional aspects have been widely undocumented. The eggs of several *Phyllium* species, including *P. philippinicum*, resemble plant seeds and bear protruding exochorionic structures (pinnae, according to Clark [43]). The morphology of these pinnae is suggested to be species-specific and their taxonomic use has been previously well demonstrated [4,42]. The functionality of these structures is thus far largely unknown, but the unfolding behavior of the pinnae is often observed in captive breeding. The fact that the pinnae morphologically respond to water has anecdotally raised questions amongst the phasmid breeding community on what purpose this mechanism might serve. Only very few taxonomic studies hypothesized the function of these pinnae. Hennemann et al. [3] described the unfolding of the pinnae after their contact with water and suggested an adhesive function of this system, however did not further elucidate this idea. Additionally curious, the oviposition technique employed by the females, does not involve active gluing of the eggs, which begs the question of whether there is a presence or absence of accessory reproductive glands in *P. philippinicum*. Unfortunately, most studies on leaf insects solely focus on external morphological features, leaving this question unanswered.

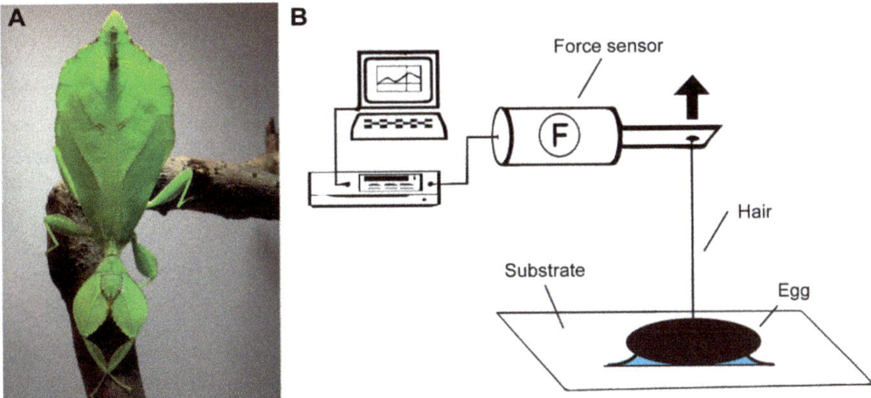

Figure 1. Examined species and experimental setup. (**A**) Female of *Phyllium philippinicum* (image is provided by Daniel Dittmar). (**B**) Experimental set-up for detachment force measurements. The egg, which was glued onto the particular substrate fixed on a lab boy using double-sided sticky tape. A hair was glued onto the egg and connected to a force sensor. To detach the egg from the substrate, the force sensor was moved away from the egg in perpendicular direction. The time force signal was amplified and finally processed in a computer.

Overall, strong egg attachment has been reported in a number of other insect species on natural substrates [44–47] and even stronger adhesion was measured from extracted egg glue on various artificial substrates [48–50]. The specific properties of the egg glue seemingly depend on the level of specialization in the attachment system and the habitat/substrate it is specialized for, therefore resulting in different strategies [51]. One important component influencing attachment efficiency is the roughness of the substrate: rougher surfaces create a greater contact area for glue and stronger adhesion after glue solidification [52]. The high complexity in the structural features of plant leaves (trichomes, wax crystals, stomata and cuticle foldings) and fruits (microcracks and epicuticular wax crystals) of various plant cultivars leads to rougher surfaces and increases the adhesion of the eggs of the codling moth *Cydia pomonella* (Linnaeus, 1758) (Lepidoptera, Tortricidae), as experimentally shown [44,45]. The egg attachment strength of the parasitic warble fly (Diptera, Hypoderminae) positively correlates to the roughness of the hairs on its host species [53]. Insect vectors of the human bot fly *Dermatobia hominis* (Linnaeus Jr., 1781) (Diptera, Oestridae) are covered with setae, which enhances egg adhesion for the human bot fly [54]. Another important factor influencing the attachment of eggs is surface chemistry. Eggs of the asparagus beetle *Crioceris asparagi* (Linnaeus, 1758) (Coleoptera, Chrysomelidae) adhere well to the surfaces of the plant *Asparagus officinalis* L. (Asparagaceae), which have superhydrophobic and microstructured surfaces due to the coating by wax crystals [46].

Eggs with adhesive responses in contact with water are only reported for a few insect species. The dragonfly *Libellula depressa* Linnaeus, 1758 (Odonata, Libellulidae), and other Anisoptera [55–64] lay eggs which possess an adhesive coating that swells and generates adhesive properties after the female deposits them in water [65]. The eggs of Ephemeroptera are covered with a thick layer composed of tightly entwined filaments, causing cohesion of the eggs and adhesion to a substrate after deposition into water [66]. The exochorionic structures of these species undergo modifications upon interaction with water, in turn generating adhesion [65,66]. It is assumed that in lieu of colleterial glands [55,56,67], these adhesive coatings are synthesized by follicle cells [65,68] which are involved in eggshell deposition [66,69–71].

On one hand, exploring the adhesive properties and response to water contact of the eggs of *P. philippinicum* can enhance our knowledge of multifunctional bioadhesives. On the other hand, this functional system can provide insights into the life history of this species and shed light on the ecological environments this species inhabits, as this knowledge is usually missing in taxonomic descriptions of museum specimens. This could assist future studies in obtaining broader ecological knowledge of this species, contributing to conservational aspects for both phasmids and plants that can be subject to damage by insects, and also give input on evolutionary studies, as the highly specialized attachment mechanism of *P. philippinicum* is highly derived. In this paper, we asked the following specific questions. (i) How do the eggs of *P. philippinicum* adhere? (ii) How do water contact, surface topography and surface chemistry influence egg adhesion in this species? (iii) Is attachment in *P. philippinicum* eggs reversible and repeatable?

2. Materials and Methods

2.1. Specimens

The eggs of *Phyllium philippinicum* Hennemann, Conle, Gottardo and Bresseel, 2009 were obtained shortly after being laid by female insects from the culture of Kirsten Weibert (Jena, Germany). The animals were fed with blackberry leaves ad libitum and kept in a natural day/night cycle. The weight of freshly laid eggs ($N = 20$) was measured using an analytical balance AG204 Delta Range microbalance (Mettler Toledo, Greifensee, Switzerland; d = 0.1 mg).

2.2. Morphology

Eggs attached to microscopy glass slides were observed with the Leica Microscope M205 (Leica Microsystems Ltd., Wetzlar, Germany). Images were captured from both sides, overview of the egg and view of the contact through the glass slide, using the microscope camera Leica DFC420 (Leica Microsystems Ltd., Wetzlar, Germany). Multifocus stacked images were postprocessed using the software Leica Application Suite (LAS) version 3.8.0 (Leica Microsystems Ltd., Wetzlar, Germany) and Affinity Photo (Apple Inc., Cupertino, CA, USA).

For higher magnification, eggs in contact with different substrates, as well as detached and untreated eggs, were air-dried and sputter-coated with gold-palladium of 10 nm thickness. The substrates corresponding to the detached eggs were sputter-coated as well. Additionally, some untreated eggs were dehydrated using an ascending alcohol series, critical point-dried and sputter-coated as well. These samples were observed in the SEM Hitachi S4800 (Hitachi High-technologies Corp., Tokio, Japan) at an acceleration voltage of 5 kV. Subsequently, the images were processed with Affinity Photo (Apple Inc., Cupertino, CA, USA).

The nomenclature of the egg morphology follows Sellick [23].

2.3. Detachment Force Measurements

The detachment force of individual eggs was measured in four different experiments. In all experiments, the eggs were mounted on standardized surfaces, as described below, and individually attached to a force transducer (100 g capacity; FORT100, World Precision Instruments Inc., Sarasota, FL, USA) by gluing a horsehair with bees wax onto the lateral side of the egg (Figure 2B) and attaching the hair to the sensor (Figure 1B). The force transducer was connected to a BIOPAC Model MP100 and a BIOPAC TCI-102 system (BIOPAC Systems, Inc., Goleta, CA, USA). Force–time curves were recorded by pulling the eggs off the surfaces using the software Acqknowledge 3.7.0 (BIOPAC Systems Inc., Goleta, CA, USA). The test surfaces were lowered away from the sensor with a speed of approximately 2–3 cm/s using a laboratory lifting platform. In all four experiments, the detachment force was measured by pulling the egg off of a surface at an angle of 90°, with the same setup, as described by Wohlfart et al. [72] for spiders and later used for adult stick insects [19]. The highest peak of the visualized graph was interpreted as the maximum detachment force. All surfaces were carefully cleaned with 70% isopropylic alcohol prior to each experiment. Detachment forces were measured in the following four different experiments:

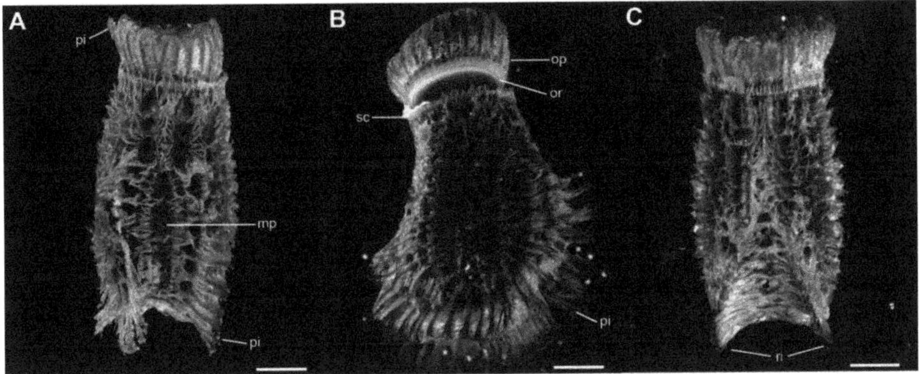

Figure 2. Morphology of the eggs of *Phyllium philippinicum*. (**A**) Dorsal view. (**B**) Lateral view. (**C**) Ventral view. **mp**, micropyle; **op**, operculum; **or**, opercular rim; **pi**, pinnae; **ri**, ribs; **sc**, serosal cuticle. Scale bars: 1 mm.

(1) Freshly laid eggs (N = 32 per substrate) were mounted on four test substrates with different roughness (0, 1 and 12 µm, and standardized p40 polishing paper) made of epoxy resin (as described below). Eggs were prepared on the test substrates by placing individual droplets of distilled water (~100 µL) on the epoxide plates and then placing one egg in a single droplet, to trigger the unfolding of the pinnae. Subsequently, the eggs were allowed to dry completely (~24 h) and then attached to the sensor.

(2) Eggs (N = 20 per substrate) were mounted on three surfaces with different chemical surface properties with the same procedure as described above. The surfaces used differed in the wettability, indicated by the contact angle of the water, which was 36.25 ± 1.15° (mean ± SD, n = 10) (hydrophilic), 83.38 ± 0.89° (the same epoxy resin as used for experiment 1) and 98.9 ± 0.47° (hydrophobic).

(3) Additionally, eggs were placed on the hydrophobic and the hydrophilic substrates in wet condition (N = 20 per substrate) and the detachment force was measured. The eggs were individually fastened with a horsehair as described above and fully submerged in distilled water for 20 min; afterwards, they were attached to the force transducer and then placed on the test substrate. After letting the eggs sit on the substrate for 1 min, the detachment force from the substrate was measured in the same manner as in the other experiments.

(4) The reproducibility of egg attachment was tested by subsequent pull-off measurements of the same egg. Individual eggs (N = 8) were prepared as described in the first experiment and attached to the smooth epoxy resin substrate (0 µm roughness). Then, the detachment force was measured by pulling off the egg. Afterwards, the same egg was then reattached once again using a droplet of water and left to dry for another 24 h. This procedure was repeated for each of the eight eggs six different times, until the measured detachment forces were similar in comparison to the previous day (i.e., revealed no significant difference).

All experiments were performed at 19–21 °C temperature and 45–55% relative humidity.

2.4. Surface Preparation

Two different types of surfaces were used in the experiments. Epoxy resin with a different surface roughness for the first and the fourth experiment and glass with different wettability, as well as epoxy resin, for the second and third experiments.

2.4.1. Glass

Clean microscope glass slides (Carl Roth GmbH & Co. KG, Karlsruhe, Germany) were used as the hydrophilic substrate and silanized, as described by Voigt and Gorb [46], to obtain a hydrophobic substrate. The surface chemistry was characterized by measuring the contact angle of the water on the substrate (aqua Millipore, droplet size = 1 µL, sessile drop method; $n = 10$ per substrate) using the contact angle measurement instrument OCAH 200 (Dataphysics Instruments GmbH, Filderstadt, Germany). The contact angle of the water was $36.25 \pm 1.15°$ for the hydrophilic glass substrate and $98.9 \pm 0.47°$ for the hydrophobic one.

2.4.2. Epoxy Resin

Substrates with different roughness were produced using epoxy resin [73] following the protocol of Salerno et al. [74]. Negative replicas were cast using polyvinylsiloxane (PVS)-based two-component dental wax (Colthéne/Whaledent AG, Altstatten, Switzerland). Negatives were then filled with epoxy resin and cured at 70 °C for 24 h. Glass (0 µm roughness) and polishing papers with the roughness of 1 µm, 12 µm (Buehler, Lake Bluff, IL, USA) and industrially standardized p40 polishing paper (particle size ~440 µm) were used as templates for the resin replicas. The contact angle of the water on the smooth epoxy resin was $83.38 \pm 0.89°$ (mean $\pm$ SD, $n = 10$).

2.5. Statistical Analysis

Statistical analyses were performed with SigmaPlot 12.0 (Systat Software Inc., San José, CA, USA). Normal distribution and homoscedasticity were tested using the Shapiro–Wilk test and Levene's test, respectively, prior to other tests. As the respective data were neither parametric nor showed homoscedasticity, detachment forces of eggs on substrates with different surface roughness, as well as on surfaces with different chemical properties represented by corresponding contact angles, were compared using Kruskal–Wallis one-way analyses of variance (ANOVA) on ranks followed by Tukey's post hoc test. Detachment forces of wet and dry eggs on surfaces with different contact angles were compared using Kruskal–Wallis one-way ANOVA and Tukey's test as well. The Mann–Whitney rank sum test was used to compare the detachment forces of eggs in the wet condition on hydrophilic and hydrophobic surfaces. For a comparison of the detachment forces over a six-day period of repeated measurements, a Friedman repeated measures ANOVA was performed along with a Tukey's post hoc test.

3. Results

3.1. Egg Morphology

The eggs of *Phyllium philippinicum* are laterally compressed and densely covered with small exochorionic appendages (pinnae, sensu Clark [43]). These pinnae cover most of the egg's surface, except for some circular pits and the center of the micropylar plate (Figure 2A). A corona of shorter expansion surrounds the micropylar plate, oriented away from it. The anterior pole of the egg is covered by an operculum, the lid of the egg, which is released during the hatching of the nymph (Figure 2B). A formation of larger pinnae surrounds the outer rim of the operculum anteriorly. Two ribs along the lateral ridges of the egg are covered with long pinnae as well, expanding the lateral dimensions of the egg (Figure 2). The ribs meet on the ventral side of the egg (Figure 2C). Pinnae of freshly laid eggs lie flat on the surface of the egg, but unfold after contact with water, as described below (Figure 3). Dimensions of the eggs are measured according to Sellick [23]. They measure 4.39 ± 0.36 mm (mean $\pm$ SD, $N = 7$) in length, with a height of 2.77 ± 0.25 mm, and width of 2.16 ± 0.14 mm. The mean weight was 15.9 ± 1.3 mg ($N = 20$).

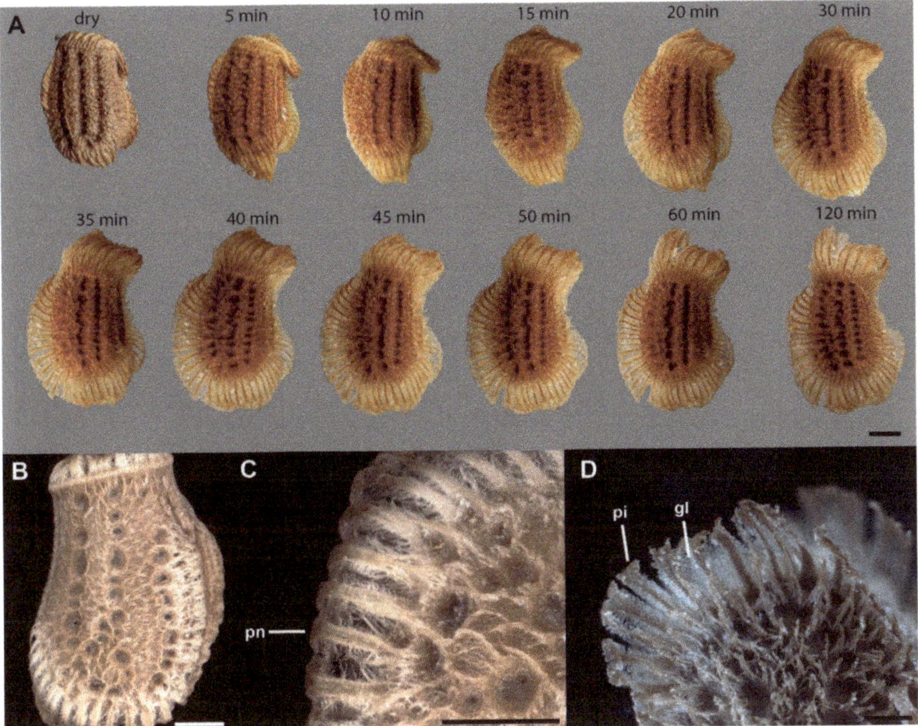

Figure 3. Unfolding behavior of *Phyllium* egg pinnae. (**A**) Succession of pinnae unfolding in *Phyllium rubrum* Cumming, Le Tirant and Teemsma, 2018, after exposure to water (images are provided by Bruno Kneubühler), lateral views. *B,C*. Lateral view of untreated *Phyllium philippinicum* egg. (**B**) Overview. (**C**) Detail of folded pinnae. (**D**) Detail of unfolded pinnae of a *Phyllium philippinicum* egg after water exposure. **gl**, glue; **pi**, unfolded pinna; **pn**, folded pinna. Scale bars: 1 mm (**A,B**), 500 µm (**C,D**).

3.2. Pinnae Behavior and Adhesive Secretion

The eggs are deposited by the female with the pinnae folded on the surface of the egg. A single pinna consists of a central shaft that is hierarchically split several times towards the tip (Figure 3D, 4). After oviposition, before initial contact with water, the folded pinnae are covered with an iridescent layer of a solidified secretion deposited by the female (Figure 3B,C). The pinnae unfold after contact with water and the secretion liquefies (Figure 3A). The larger pinnae on the operculum and the lateral ribs of the egg unroll and expand the dimension of the projected lateral area of the egg. Smaller pinnae, as well as hierarchical expansions of the main fringes of the pinnae, expand and increase the egg surface as well. The liquefied secretion on the surface of the eggs, after expansion of the pinnae, concentrates on the tips of the expansions (Figure 3D). Along the length of larger pinnae, a reservoir of the secretion forms a bridging film between the shafts of the pinnae. During contact with a substrate, the pinnae deform and spread the viscous secretion, in which they are imbedded, onto the substrate. After some time without contact to water (5–6 h), the secretion dehydrates and solidifies again (Figure 4A,B). After the curing off the secretion, the egg remains attached to the substrate. The adhesive function of the glue is characterized below.

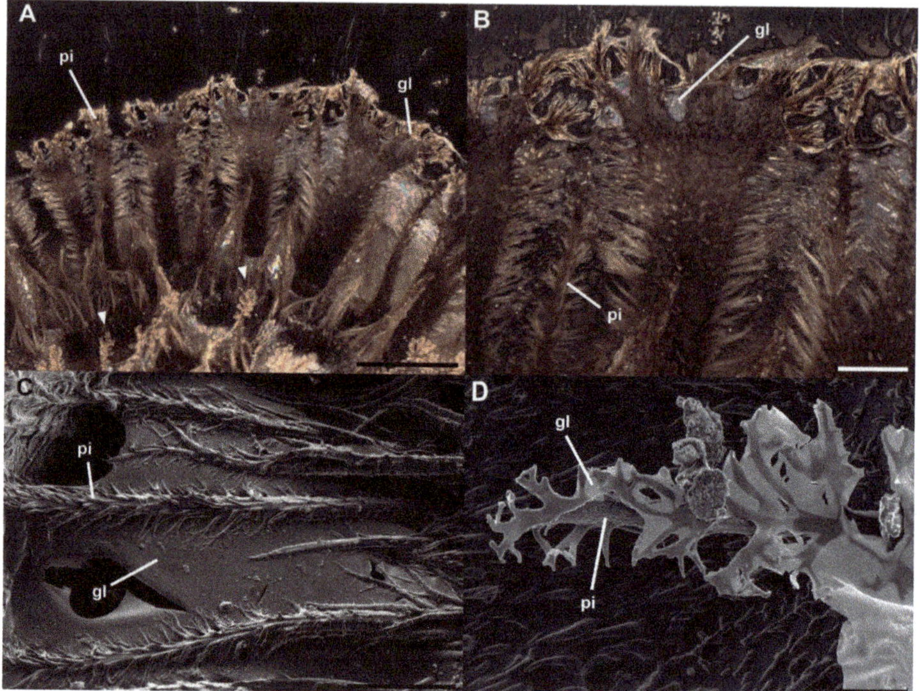

Figure 4. Glue associated with *Phyllium philippinicum* pinnae. (**A,B**) Stereomicroscopic images of pinnae attached to a glass surface, view through the glass slide. (**A**) Glue deposition on a glass surface and pinnae interaction with the substrate (arrowheads). (**B**). Reinforcement and distribution of the glue by the pinnae. (**C,D**) Scanning electron microscopy images of glue–pinnae interactions. C. Glue film adhering to pinnae. (**D**) Dried glue residuals on a pinna after detachment from a smooth glass surface. **gl**, glue; **pi**, pinnae. Scale bars: 500 μm (**A**), 300 μm (**B**), 100 μm (**C**), 10 μm (**D**).

3.3. Egg Attachment

The attachment performance of *P. philippinicum* eggs on different surface roughnesses is illustrated in Figure 5. The maximum pull-off force measured before the egg detached from the respective substrate (maximum detachment force, Figure 5A) is considered a measure for the attachment capability of the egg to the substrate. The maximum detachment force values were highest on the intermediate roughnesses, 12 μm with 144.65 ± 133.38 mN (median ± SD) and 1 μm with 144.23 ± 137.18 mN. The lowest detachment forces were recorded on the roughest (p40; 81.71 ± 104.11 mN) and the smoothest (0 μm; 122.94 ± 95.28 mN) surfaces. However, the differences in median detachment force values between the four surface roughnesses were not significant (Kruskal–Wallis one-way analysis of variance (ANOVA), $H = 7.278$, d.f. $= 3$, $p = 0.064$, $N = 32$ per roughness).

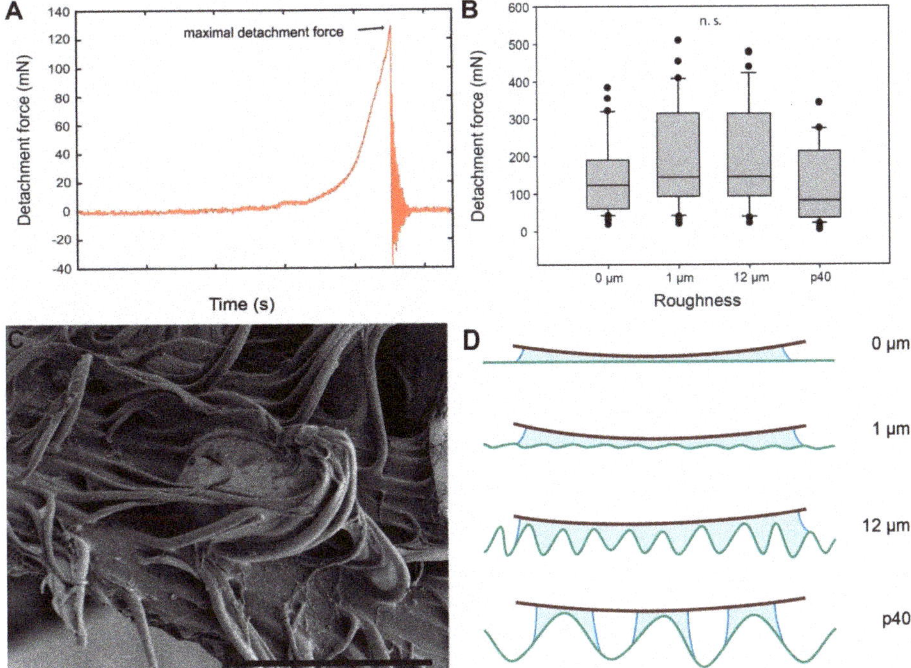

Figure 5. Influence of roughness on egg adhesion. (**A**). Exemplary force-time curve from measurements of the detachment force. (**B**) Detachment forces from substrates with different surface asperity ($N = 32$ for each roughness). Boxes are 25th and 75th percentiles, the line within the boxes defines the median, and whiskers represent the 10th and 90th percentiles. **n. s.** = no statistical difference ($p > 0.05$, Kruskal–Wallis ANOVA on ranks). (**C**) Scanning electron microscopy image of pinnae deformation showing the adaptation of pinna extensions to surface corrugations. (**D**) Schematic interpretation of the eggs' glue with differing degrees of surface roughness. Roughness parameters are given in detail by Salerno et al. [74]. Scale bar: 60 μm.

The attachment performance of eggs on surfaces of differing surface chemistry is displayed in Figure 6A. The detachment force from pulling the eggs off of the hydrophilic surface (water contact angle 36.25°) was very high (792.37 ± 293.94 mN) and significantly higher than the force measured on surfaces with a higher water contact angle (Kruskal–Wallis one-way ANOVA, $H = 38.543$, d.f. = 2, $p \leq 0.001$, $N = 20$ per surface; Tukey's test, $p < 0.05$). The adhesion to epoxy resin (water contact angle 83.38°) was significantly lower than that of the hydrophilic glass with 159.03 ± 117.31 mN (Tukey's test, $p < 0.05$), but higher than the adhesion to the hydrophobic glass (water contact angle 98.9 °) with 88.03 ± 114.81 mN. The latter difference, between the epoxy resin and hydrophobic silanized glass, was not found to be statistically significant according to Tukey's post hoc test (Tukey's test, $p > 0.05$).

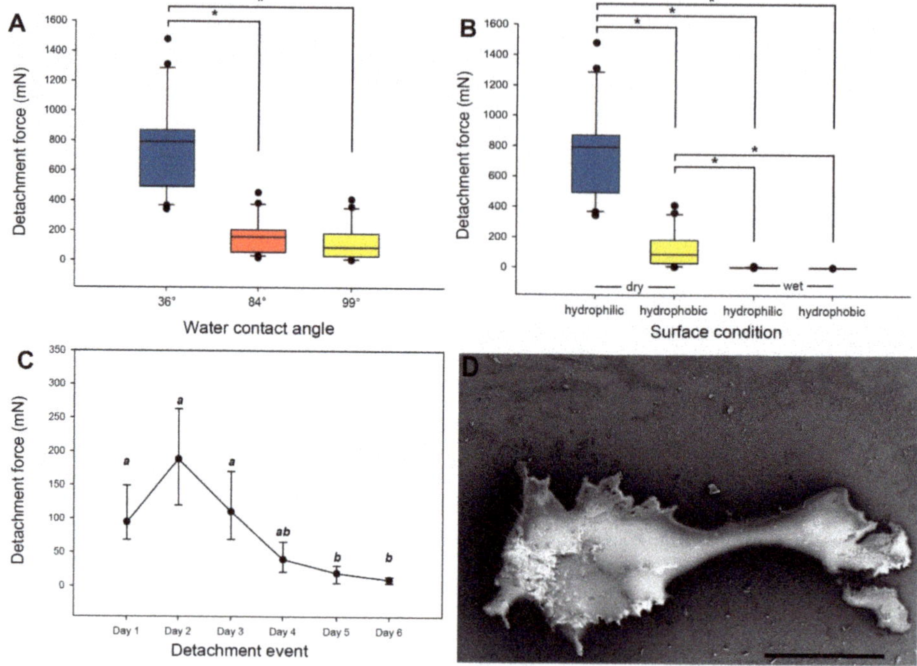

Figure 6. Influence of surface chemistry and repetitive detachment on *Phyllium philippinicum* eggs. (**A**) Detachment forces from surfaces with different water contact angles ($N = 20$ for each contact angle). (**B**) Detachment forces from wet and dry surfaces with different chemical properties ($N = 20$ for each treatment). "Hydrophilic" corresponds to a contact angle of 36° and "hydrophobic" corresponds to a 99° water contact angle. Boxes are 25th and 75th percentiles, the line within the boxes defines the median, and whiskers represent the 10th and 90th percentiles. * $p \leq 0.001$, only significant comparisons are highlighted (Kruskal–Wallis one-way ANOVA on ranks followed by Tukey's post hoc test). (**C**) Detachment forces during sequential detachment events ($N = 8$ eggs). Dots indicate the median, whiskers represent the standard deviation. Lowercase letters indicate statistically similarity. Groups with the same letter are statistically equal (Friedman repeated measurements ANOVA, followed by Tukey's test). (**D**) Scanning electron microscopy image of glue residuals on a smooth hydrophobic glass surface. Scale bar: 300 µm.

Adhesion to both hydrophobic and hydrophilic surfaces was very low and practically negligible in the presence of water on the surface (Figure 6B). Wet eggs on the hydrophobic surface (2.74 ± 0.34 mN) showed no significant difference in detachment values compared with wet eggs on the hydrophilic surface (3.17 ± 2.30 mN; Kruskal–Wallis one-way ANOVA, $H = 66.77$, d.f. = 3, $p \leq 0.001$, $N = 20$ per treatment; Tukey's post hoc test, $p > 0.05$). The comparison of the egg adhesion performance between hydrophobic and hydrophilic surfaces in both wet and dry conditions yielded significant differences between all comparisons, except for the comparison between the two wet surfaces ($p < 0.05$, Tukey's test). Eggs dried in adherence to hydrophilic surfaces showed significantly higher detachment forces than eggs in contact with wet surfaces, as well as higher adhesion than eggs dried on hydrophobic surfaces (all $p < 0.05$, Tukey's test). The detachment force of dried eggs from the hydrophobic glass was lower than from the hydrophilic substrates, but higher than wet eggs from both substrates (all $p < 0.05$, Tukey's test).

Figure 6C illustrates the attachment performance of eggs over a sequence of six repeated detachment events. The median detachment force initially increased from day 1 (94.8 ± 38.0 mN) to day 2 (188.68 ± 66.75 mN). Subsequently, the detachment force consistently decreased from day 2 until day 6 (9.79 ± 4.56 mN). The detachment force was statistically different (Friedman repeated measures ANOVA on ranks, $\chi^2 = 35.179$, d.f. $= 5$, $p \leq 0.001$, $N = 8$ per day) and decreased between day 2 and day 6. However, the first three days were statistically similar, but each of the days 1–3 were significantly higher than days 5 and 6 (all $p < 0.05$, Tukey's test). Although the overall decrease in attachment performance is significant, the initially higher median detachment force on day 2 is not significantly different from day 1 and day 3 ($p > 0.05$, Tukey's test).

4. Discussion

4.1. Attachment Mechanism

The attachment capabilities of the eggs of *Phyllium philippinicum* were not readily paid attention to in the past and recognized only anecdotally in the literature [3]. However, the combination of an adhesive secretion and reinforcing microstructured exochorionic structures has proven to provide excellent attachment. The mean safety factor (F_a/F_m; mean detachment force per weight force) of eggs on a smooth epoxy resin substrate ranges around 924, i.e., the adhesion of one egg sufficiently attaches 924 times its own weight. On hydrophilic substrates, the average F_a/F_m is 4825.

Water exposure has two main effects on the egg: (1) unfolding of the pinnae and (2) liquefaction of the glue (Figure 3). Both effects contribute to the enhancement of the adhesive properties of the eggs. Like a solvent-based adhesive, the egg adhesive dissolves partially in water and once the water evaporates, the adhesive dries and hardens on the substrate. When introduced to water, the pinnae extend and fan out, adapting to the texture of the substrate. The liquid glue covers the pinnae, which transmit and spread it out onto the substrate. Such horizontally oriented fibrillary structures, that lay parallel to a surface, facilitate the spreading of a fluid, hence enhancing the surface contact of the adhesive fluid [75]. Therefore, bridges of dried adhesive material between adjacent pinnae are visible, when re-solidified (Figure 3D, 4C). To achieve proper attachment, the glue becomes fluid to interact with an adjoining surface, then the adhesive fluid dries to either create a sufficient contact area [76,77] at the interface or mechanically interlock with the surface irregularities [77,78]. Whether high humidity in the surroundings or solely the contact to water droplets cause the glue to liquefy remains untested. To evaluate the effect of ambient humidity, further experiments with exposure to differing humidity are necessary. While the pinnae facilitate adhesion through an increased contact area with the surface, the fluid adhesive makes large real contact with the surface. The exochorionic extensions may also be able to extend into and interdigitate with surface asperities and further spread the adhesive fluid [79–81], depending on the roughness profile of the surface (Figure 5C). Their hierarchical structure offers finer subcontacts with the substrate [82,83], and, hence, optimizes contact formation on natural surfaces of fractal roughness with overlapping wavelengths (e.g., tree bark) [84]. Overall, the pinnae reinforce the film of the glue, thus achieving a viable adhesive system: soft enough to form intimate contact, yet stiff enough after solidification, to decrease elastic deformation and hold a strong bond [84]. This pinnae-based reinforcement offers structural integrity to the adhesive system of the egg.

Besides the mechanical interlocking of the solidified glue with surface corrugations, the glue adheres by physiochemical interactions, presumably van der Waals that can prove very strong with sufficient interfacial contact [77].

4.2. Influence of Substrate Roughness

Attachment on substrates with different surface roughness revealed no significant differences among all tested surfaces (Figure 5B). Other biological attachment systems are found to be significantly affected by surface roughness. The tarsal attachment systems of some flies and beetles consist of tenent setae that, similar to the pinnae of *P. philippinicum* eggs, adapt to the surface profile [85–91]. However, the performance of these attachment systems for the purpose of locomotion fare better on smooth surfaces or rougher surfaces exceeding asperity sizes of 3 μm, with the worst performance on micro-roughnesses ranging from 0.1–0.3 μm. This is explained by the spatula-like terminal elements of insects' tenent setae interaction with the surface [85,89,90]. These setae tips are able to make sufficient contact with large surface asperities but are confounded by micro-rough surfaces that inhibit real contact of the setae to the surface. The eggs of *P. philippicum*, in contrast, performed well on all surface roughnesses tested.

This ability of *P. philippicum* eggs is presumably based on the action of initially fluid and later solidified glue. For glues, like that of the adhesive material of the eggs examined herein, rougher surfaces create a larger contact area and stronger adhesion [52]. This applies to the performance of *P. philippinicum* eggs, with an increasing trend in attachment strength from 0–12 μm roughness. Adhesion relies on the area of actual contact made with a surface [79–81]. Although surfaces with micro-roughness are not generally favourable for many insect attachment systems associated with locomotion, the egg's adhesive fluid is able to conform around small surface irregularities and hence increase the actual contact area (Figure 5D). Rough surfaces are beneficial for egg adhesion in different insect species. For the codling moth, it has been previously shown that smoother surfaces with fewer trichomes and rather low free surface energy deter their eggs' attachment [44], while structural features creating a rougher surface on leaves or fruits (e.g., trichomes, microcracks or epicutilar wax crystals) lead to stronger attachment of codling moth eggs due to an increase in the contact area with the egg's glue [45]. Rough surfaces on plants are known to be favorable choices for oviposition sites in other lepidopterans [92,93] and the willow leaf beetle *Phratora vulgatissima* (Linnaeus, 1758) (Coleoptera, Chrysomelidae) [94], leading to an enhanced attachment strength. The same applies for the surface texture of the oviposition substrates of parasitic flies [53,54]. In contrast, the eggs of *P. philippinicum* adhere similarly strong to surfaces of all tested roughnesses. As the eggs are dropped without direct oviposition onto specific substrates, this case of universal adhesion is most likely adapted to a broad spectrum of surface roughness.

However, p40, the roughest surface tested, revealed the weakest overall attachment to the eggs. The adhesive fluid probably generates a larger contact area on substrates with micro-rough surface corrugations. The large surface asperities of the p40 substrate (~440 μm asperity size) principally offer a larger surface for contact formation, but the glue of the egg presumably does not fill the deeper asperities, creating only partial interaction with the walls of the surface asperities (Figure 5D). The viscosity of the glue presumably prevents the glue from properly filling the surface texture.

4.3. Influence of Surface Chemistry

The adhesive strengths on surfaces with different water contact angles revealed significant differences between the attachment strength of eggs on the hydrophilic substrate compared with the two substrates with higher water contact angles. The contact formation and generation of attachment by the adhesive fluid depends on the surfaces' chemical properties. Higher free surface energy and lower contact angles, which essentially vary inversely to one another [95], are characteristics of surface hydrophilicity that invite greater wetting of liquids on such a surface, in turn forming greater contact at the liquid–surface interface, which creates stronger adhesion [82,96,97]. Lower surface energy reduces the overall attachment ability of a system [84] and therefore a lower surface energy results in a lower detachment force of the eggs tested on the hydrophobic substrate. The same correlation between surface chemistry and attachment is reported for the tarsal attachment system of several groups of insects [87,88,98,99]. The water-mobilized adhesive presumably does not wet hydrophobic surfaces properly and therefore attachment is reduced, as wetting is an important prerequisite for this type of adhesion [78,100]. The adhesive fluid and its composition presumably best perform on hydrophilic surfaces, most likely due to polarity within the adhesive fluid and water that is attracted to polar/hydrophilic surfaces [78].

The range of suitable surface chemistry regimes in insect attachment is presumably a result of co-evolution of the insects and their corresponding plants [95]. Adaptation to substrate chemistry is species-specific and depends on the degrees of specialization to various natural substrates. Unfortunately, not many aspects of the ecology and natural habitat of *P. philippinicum* are known. Therefore, assumptions about their host trees are based on diet compatibility with certain leaves from their endemic region [3] and known host species of closely related *Phyllium* [1,101]. As *P. philippinicum* are generalist phytophages, there are several potential host plants. Some of the supposed host species of *P. philippinicum* are *Psidium guajava* L. (Myrtaceae), *Mangifera indica* L. (Anacardiaceae) and *Nephelium lappaceum* L. (Sapindaceae) [1,3,101]. The higher attachment performance on hydrophilic surfaces does not allow to specifically determine the actual host plants but enables an approximation of natural substrates the eggs are adapted to. All three putative food plants are evergreen tree species which have leaves that are generally hydrophobic with water contact angles around 100° [102]. In contrast, the bark of guava and mango was estimated to have contact angles around 52° and 50° [103]. The eggs more likely adhere to the bark than to the leaves. This is also reflected in the coloration of the eggs and freshly hatched nymphs. Nymphs hatch a dark brown hue and change to green after feeding on the foliage of the tree. Further tests are certainly necessary to test this assumption, but based on the findings herein, *P. philippinicum*'s eggs have a better chance at attaching to wood surfaces of its host tree species compared with leaf surfaces. Additionally, anchorage of the egg to a substrate could help the new-born nymph successfully hatch from the inside of the egg while remaining arboreal.

Apparently, for the eggs of *P. philippinicum*, the surface chemistry influences egg adhesion more than surface roughness. In experiments with beetles, their tarsal attachment systems worked the other way around. The hairy, wet adhesive pads of *Coccinella septempunctata* Linnaeus, 1758 (Coleoptera: Coccinellidae) and *Leptinotarsa decemlineata* Say, 1824 (Coleoptera: Chrysomelidae) responded more to surface roughness than to surface chemistry [51,104]. However, *Gastrophysa. viridula* (De Geer, 1775) (Coleoptera: Chrysomelidae) exhibited a decreasing attachment performance with hydrophobicity and stronger performance on smooth surfaces [87]. On the other hand, the eggs of *P. philippinicum* show decreased attachment on hydrophobic surfaces, but attachment values were high for all surface roughnesses. However, for attachment of the eggs, in contrast to moving insects, the preconditions are different.

Two studies on egg adhesion of the codling moth used various natural surfaces with specific topographies and physiochemistries [44,45], however, the glue of the codling moth is permanent and most likely water-insoluble, and the adhesive of *P. philippinicum* is reversible and mostly water-soluble. Furthermore, these studies also found an effect of surface topography, which is probably due to the specific topography of the natural surfaces. Regardless, further investigation into the combined effect of surface chemistry and topography using a wider scope for both would be advantageous for the comparative effect on the attachment of *P. philippinicum* eggs.

4.4. Glue Properties

Detachment force values from *P. philippinicum* eggs in the wet condition were extremely low and practically negligible in comparison with other force measurements. It is assumed that the low force values in the wet condition reflects surface tension and capillary forces exerted by the water droplet that the egg was submerged in upon detachment, not the eggs' actual attachment to the surface.

Many insect eggs require hydration to become adhesive [55,57–66]. For example, eggs of the mayfly *Siphlonurus lacustris* (Eaton, 1870) (Ephemeroptera: Siphlonuridae) have a thick fibrous coat surrounding their eggs that undergo exochorionic changes once deposited in water, creating cohesion between egg masses and adhesion to a substrate [66]. However, these eggs attach in the wet condition and remain underwater. Contrastingly, the eggs examined herein need to dry after hydration to generate adhesion. The egg of *P. philippinicum* does not achieve attachment in water and necessitates a phase change from liquid to solid for the adhesive material to adhere. This could serve as a mechanism for the optimum site selection for the incubation of the egg to avoid adhesion under water, as attachment under water would be lethal to the hatching nymph. The eggs of *P. philippinicum* may be almost immediately ready to attach to a substrate once produced from the mother due to high humidity and the prevalence of water in a tropical rainforest [105]. This makes sufficient hydration of the egg highly probable and triggers its adhesive capabilities. However, if suitable conditions are not found, the egg retains the potential to adhere in a suitable environment in the next attachment event.

The glue mediated attachment of *P. philippinicum* eggs is reversible and reproducible over several cycles of attachment, detachment and reattachment (Figure 6C). Apparently, no glue is secreted by the egg itself; furthermore, colleterial glands for glue production are probably absent in the females as well [106], similar to some odonates and mayflies [65,66]. The fact that the attachment strength decreases after a few cycles of reattachment is probably due to the depletion in the supply of the glue covering the egg. The test surfaces revealed residuals of the glue after detachment (Figure 6D). However, the repeated ability to attach after submersion in water shows that the adhesive material is not entirely water-soluble and most of the secretion remains on the egg. In all likelihood, a hydrophilic polar portion of the material allows diffusion in water, facilitating adsorption at the glue interface [107] and consequently facilitates contact adaptation to the substrate [78]. A hydrophobic nonpolar portion most likely remains on the egg, preventing full dissolution [108]. The oviposition method that *P. philippinicum* females employ does not give them as much control over the oviposition site compared with other species that directly deposit eggs. The reversibility of adhesion may be a technique to correct maladaptive attachment sites or to adapt to seasonal changes in the environment, as reported for the egg glue of some alpine butterflies [47].

The chemical composition of the glue remains ambiguous and is not the subject of this study. However, some assumptions can be drawn based on the experimental results. Most permanent bioadhesives involved in egg attachment are largely proteinaceous [48,109–115]. The amphiphilic nature of the glue could be achieved by glycoproteins, as in many other insect glues [116]. The highly soluble glycan would serve as the polar portion [48,114,117], facilitating non-covalent bonding with hydrophilic substrates. The protein serves as the hydrophobic portion [108,118], providing adherence of the glue to the surface of the egg and its appendages.

5. Conclusions

Although the special appearance of the eggs of *Phyllium* species, including *P. philippinicum*, received attention in different taxonomic and evolutionary studies [15,23,42], only a few hypotheses on the function of the special morphological features were presented [3]. We herein elucidate the specialized attachment mechanism of the eggs of this species and provide the first experimental approach to systematically characterize the functional properties of their adhesion. The adhesive mechanism of the egg exploits a combination of folded exochorionic structures (pinnae) and a film of adhesive secretion. Both components respond to contact with water. The glue becomes fluid through hydration, adapts to the substrate profile and adheres after solidification. The pinnae facilitate the spreading of the glue, support adaptability using hierarchically splitting filaments and reinforce the hardened film. This mechanism copes with surface roughness using this combination but is affected by surface chemistry. The glue adheres very well to hydrophilic surfaces, but the attachment force decreases with an increasing water contact angle. Although the egg cannot achieve attachment while submerged in water, it can reattach itself after dislodgement from a surface, making its adhesive mechanism temporary, and arguably long-term [77], depending on the conditions. This replicability of attachment can accomplish attachment site optimization to ensure suitable environmental conditions for embryonic development. This includes fixation in preferable environmental conditions, but also adjustment in case of environmental changes. The mechanism described herein copes with different degrees of surface roughness but is affected by the surface chemistry of the substrate. Other adhesive secretions in insects interestingly perform differently, although they serve a very similar function: the larval glue of the fly *Drosophila melanogaster* consists of glycosylated proteins and is used to anchor the pupa to different substrates [119]. In contrast to the egg glue of *P. philippinicum*, the glue of *D. melanogaster* larvae adheres well to various substrates independent of their surface chemistry or roughness [120].

This leads to the assumption that the egg deposition in *P. philippinicum* favors hydrophilic substrates and suggests preferable deposition site selection for the eggs in the natural habitat.

Knowledge about this mechanism can support ecologists and conservationists. Elucidating the nature of the attachment mechanism helps in understanding the dispersal as well as the life history of the species. This can help in quantifying fecundity for conservation purposes of the insect species [120]. Information on the attachment sites can help the conservation of plants and gauging the population density [1,121]. The details of this potential transitory state between non-adhesive and permanently attached eggs can be useful for evolutionary biologists.

Furthermore, this repeatable and water-controlled mechanism can stimulate biomimetic research in the field of bioadhesives [48,122–124]. The origin and biochemical nature of the glue, however, remain elusive and should be subject to future studies.

Author Contributions: Conceptualization, T.H.B. and S.N.G.; methodology, T.H.B. and S.N.G.; validation, T.H.B. and S.N.G., formal analysis, T.H.B. and E.Q.; investigation, T.H.B. and E.Q.; resources, S.N.G.; data curation, T.H.B.; writing—original draft preparation, T.H.B.; writing—review and editing, S.N.G., E.Q. and T.H.B.; visualization, T.H.B. and E.Q.; supervision, T.H.B. and S.N.G.; project administration, S.N.G.; funding acquisition, S.N.G. All authors have read and agreed to the published version of the manuscript.

Funding: This research was partially funded by the German Science Foundation, DFG grant GO 995/34-1.

Acknowledgments: We thank Jan Michels (Department of Functional Morphology and Biomechanics, Kiel University, Germany) for support in microscopy techniques. Kirsten Weibert (Jena, Germany) is thanked for providing specimens for this study, Bruno Kneubühler (Zurich, Switzerland) and Daniel Dittmar (Berlin, Germany) for providing images. We acknowledge financial support by DFG within the funding programme Open Access Publizieren. E.Q. was financially supported by the European Commission through the program Erasmus Mundus Master Course—International Master in Applied Ecology" (EMMC-IMAE) (FPA 2023-0224/532524-1-FR-2012-1-ERA MUNDUS-EMMC). Coordination F.-J. Richard, Université de Poitiers, France.

Conflicts of Interest: The authors declare no conflict of interest. The funders had no role in the design of the study; in the collection, analyses, or interpretation of data; in the writing of the manuscript, or in the decision to publish the results.

References

1. Bedford, G.O. Biology and ecology of the Phasmatodea. *Annu. Rev. Entomol.* **1978**, *23*, 125–149. [CrossRef]
2. Robertson, J.A.; Bradler, S.; Whiting, M.F. Evolution of oviposition techniques in stick and leaf insects (Phasmatodea). *Front. Ecol. Evol.* **2018**, *6*, 216. [CrossRef]
3. Hennemann, F.H.; Conle, O.V.; Gottardo, M.; Bresseel, J. On certain species of the genus *Phyllium* Illiger, 1798, with proposals for an intra-generic systematization and the descriptions of five new species from the Philippines and Palawan (Phasmatodea: Phylliidae: Phylliinae: Phylliini). *Zootaxa* **2009**, *2322*, 1–83. [CrossRef]
4. Cumming, R.T.; Bank, S.; Le Tirant, S.; Bradler, S. Notes on the leaf insects of the genus *Phyllium* of Sumatra and Java, Indonesia, including the description of two new species with purple coxae (Phasmatodea, Phylliidae). *ZooKeys* **2020**, *913*, 89. [CrossRef]
5. Büscher, T.H.; Grohmann, C.; Bradler, S.; Gorb, S.N. Tarsal attachment pads in Phasmatodea (Hexapoda: Insecta). *Zoologica* **2019**, *164*, 1–94.
6. Wedmann, S.; Bradler, S.; Rust, J. The first fossil leaf insect: 47 million years of specialized cryptic morphology and behavior. *Proc. Natl. Acad. Sci. USA* **2007**, *104*, 565–569. [CrossRef]
7. Wang, M.; Béthoux, O.; Bradler, S.; Jacques, F.M.B.; Cui, Y.; Ren, D. Under cover at pre-angiosperm times: A cloaked phasmatodean insect from the Early Cretaceous Jehol biota. *PLoS ONE* **2014**, *9*, e91290. [CrossRef]
8. Buckley, T.R.; Attanayake, D.; Bradler, S. Extreme convergence in stick insect evolution: Phylogenetic placement of the Lord Howe Island tree lobster. *Proc. R. Soc. B* **2009**, *276*, 1055–1062. [CrossRef]
9. Bell, C.D.; Soltis, D.E.; Soltis, P.S. The age and diversification of the angiosperms re-revisited. *Am. J. Bot.* **2010**, *97*, 1296–1303. [CrossRef]
10. Magallón, S.; Castillo, A. Angiosperm diversification through time. *Am. J. Bot.* **2009**, *96*, 349–365. [CrossRef]
11. Buckley, T.R.; Attanayake, D.; Nylander, J.A.A.; Bradler, S. The phylogenetic placement and biogeographical origins of the New Zealand stick insects (Phasmatodea). *Syst. Entomol.* **2010**, *35*, 207–225. [CrossRef]

12. Bradler, S.; Cliquennois, N.; Buckley, T.R. Single origin of the Mascarene stick insects: Ancient radiation on sunken islands? *BMC Evol. Biol.* **2015**, *15*, 196. [CrossRef] [PubMed]

13. Goldberg, J.; Bresseel, J.; Constant, J.; Kneubühler, B.; Leubner, F.; Michalik, P.; Bradler, S. Extreme convergence in egg-laying strategy across insect orders. *Sci. Rep.* **2015**, *5*, 7825. [CrossRef] [PubMed]

14. Simon, S.; Letsch, H.; Bank, S.; Buckley, T.R.; Donath, A.; Liu, S.; Machida, R.; Meusemann, K.; Misof, B.; Podsiadlowski, L.; et al. Old World and New World Phasmatodea: Phylogenomics resolve the evolutionary history of stick and leaf insects. *Front. Ecol. Evol.* **2019**, *7*, 345. [CrossRef]

15. Sellick, J. The range of egg capsule morphology within the phasmatodea and its relevance to the taxonomy of the order. *Ital. J. Zool.* **1997**, *64*, 97–104. [CrossRef]

16. Bradler, S. The Phasmatodea Tree of Life: Surprising facts and open questions in the evolution of stick and leaf insects. *Entomol. Heute* **2015**, *27*, 1–23.

17. Bußhardt, P.; Wolf, H.; Gorb, S.N. Adhesive and frictional properties of tarsal attachment pads in two species of stick insects (Phasmatodea) with smooth and nubby euplantulae. *Zoology* **2012**, *115*, 135–141. [CrossRef]

18. Büscher, T.H.; Gorb, S.N. Subdivision of the neotropical Prisopodinae Brunner von Wattenwyl, 1893 based on features of tarsal attachment pads (Insecta, Phasmatodea). *ZooKeys* **2017**, *645*, 1–11. [CrossRef]

19. Büscher, T.H.; Gorb, S.N. Complementary effect of attachment devices in stick insects (Phasmatodea). *J. Exp. Biol.* **2019**, *222*. [CrossRef]

20. Büscher, T.H.; Buckley, T.R.; Grohmann, C.; Gorb, S.N.; Bradler, S. The evolution of tarsal adhesive microstructures in stick and leaf insects (Phasmatodea). *Front. Ecol. Evol.* **2018**, *6*, 69. [CrossRef]

21. Büscher, T.H.; Kryuchkov, M.; Katanaev, V.L.; Gorb, S.N. Versatility of Turing patterns potentiates rapid evolution in tarsal attachment microstructures of stick and leaf insects (Phasmatodea). *J. R. Soc. Interface* **2018**, *15*, 20180281. [CrossRef] [PubMed]

22. Carlberg, U. A review of the different types of egglaying in the Phasmida in relation to the shape of the eggs and with a discussion on their taxonomic importance (Insecta). *Biol. Zentralblatt* **1983**, *102*, 587–602.

23. Sellick, J.T.C. Descriptive terminology of the phasmid egg capsule, with an extended key to the phasmid genera based on egg structure. *Syst. Entomol.* **1997**, *22*, 97–122. [CrossRef]

24. Bradler, S.; Buckley, T.R. Biodiversity of Phasmatodea. In *Insect Biodiversity: Science and Society II*; Foottit, R.G., Adler, P.H., Eds.; Wiley: Hoboken, NJ, USA, 2018; pp. 281–313. [CrossRef]

25. Moore, P.D. How to get carried away. *Nature* **1993**, *361*, 304–305. [CrossRef]

26. Stanton, A.O.; Dias, D.A.; O'Hanlon, J.C. Egg dispersal in the Phasmatodea: Convergence in chemical signalling strategies between plants and animals? *J. Chem. Ecol.* **2015**, *41*, 689–695. [CrossRef]

27. Compton, S.G.; Ware, A.B. Ants disperse the elaiosome-bearing eggs of an african stick insect. *Psyche* **1991**, *98*, 207–213. [CrossRef]

28. Hughes, L.; Westoby, M. Capitula on stick insects and elaiosomes on seeds: Convergent adaptations for burial by ants. *Funct. Ecol.* **1992**, *6*, 642–648. [CrossRef]

29. Windsor, D.M.; Trapnell, D.W.; Amat, G. The egg capitulum of a Neotropical walkingstick, *Calynda biscuspis*, induces aboveground egg dispersal by the ponerine ant, *Ectatomma ruidum*. *J. Inst. Behav.* **1996**, *9*, 353–367. [CrossRef]

30. Traveset, A.; Robertson, A.W.; Rodríguez-Pérez, J. A review on the role of endozoochory in seed germination. In *Seed Dispersal: Theory and Its Application in a Changing World*; Dennis, A.J., Schupp, E.W., Green, R.J., Westcott, D.A., Eds.; CABI: Wallingford, UK, 2007; pp. 78–103. [CrossRef]

31. Kreitschitz, A.; Kovalev, A.E.; Gorb, S.N. Slipping vs sticking: Water-dependent adhesive and frictional properties of *Linum usitatissimum* L. seed mucilaginous envelope and its biological significance. *Acta Biomater.* **2015**, *17*, 152–159. [CrossRef]

32. Shelomi, M. Phasmid eggs do not survive digestion by Quails and Chickens. *J. Orthoptera Res.* **2011**, *20*, 159–162. [CrossRef]

33. Suetsugu, K.; Funaki, S.; Takahashi, A.; Ito, K.; Yokoyama, T. Potential role of bird predation in the dispersal of otherwise flightless stick insects. *Ecology* **2018**, *99*, 1504–1506. [CrossRef] [PubMed]

34. Wang, C.-H.; Chu, Y.-I. The morphological study of the egg shell of the Tsuda's giant stick insect *Megacrania alpheus* Westwood. *NTU Phytopathol. Entomol.* **1982**, *9*, 98–109.

35. Ushirokita, M. Eggs of stick insect drifting in the wake of screw pine's seed. *Insectarium* **1998**, *35*, 108–115.

36. Kobayashi, S.; Usui, R.; Nomoto, K.; Ushirokita, M.; Denda, T.; Izawa, M. Does egg dispersal occur via the ocean in the stick insect *Megacrania tsudai* (Phasmida: Phasmatidae)? *Ecol. Res.* **2014**, *29*, 1025–1032. [CrossRef]

37. Kobayashi, S.; Usui, R.; Nomoto, K.; Ushirokita, M.; Denda, T.; Izawa, M. Population dynamics and the effects of temperature on the eggs of the seawater-dispersed stick insect *Megacrania tsudai* (Phasmida: Phasmatidae). *Zool. Stud.* **2016**, *55*, 20. [CrossRef]

38. Dehgan, B.; Yuen, C.K.K.H. Seed morphology in relation to dispersal, evolution; propagation of *Cycas*, L. *Bot. Gaz.* **1983**, *144*, 412–418. [CrossRef]

39. Nakanishi, H. Dispersal ecology of the maritime plants in the Ryukyu Islands, Japan. *Ecol. Res.* **1988**, *3*, 163–173. [CrossRef]

40. Cumming, R.T.; Leong, J.V.; Lohman, D.J. Leaf insects from Luzon, Philippines, with descriptions of four new species, the new genus *Pseudomicrophyllium*, and redescription of *Phyllium (Phyllium) geryon* Gray, 1843,(Phasmida: Phylliidae). *Zootaxa* **2017**, *4365*, 101–131. [CrossRef]

41. Cumming, R.T. A new species of *Phyllium (Phyllium)* Illiger, 1798 from Mindanao, Philippines (Phasmida, Phylliidae). *Zootaxa* **2017**, *4303*, 297–300. [CrossRef]

42. Cumming, R.T.; Le Tirant, S.; Hennemann, F.H. A new leaf insect from Obi Island (Wallacea, Indonesia) and description of a new subgenus within *Phyllium* Illiger, 1798 (Phasmatodea: Phylliidae: Phylliinae). *Faunitaxys* **2019**, *7*, 1–9.

43. Clark, J.T. The eggs of leaf insects (Insecta: Phasmida). *Zool. J. Linn. Soc.* **1978**, *63*, 249–258. [CrossRef]

44. Al Bitar, L.; Gorb, S.N.; Zebitz, C.P.W.; Voigt, D. Egg adhesion of the codling moth *Cydia pomonella* L. (Lepidoptera, Tortricidae) to various substrates: I. Leaf surfaces of different apple cultivars. *Arthropod Plant Interact.* **2012**, *6*, 471–488. [CrossRef]

45. Al Bitar, L.; Gorb, S.N.; Zebitz, C.P.W.; Voigt, D. Egg adhesion of the codling moth *Cydia pomonella* L. (Lepidoptera, Tortricidae) to various substrates: II. Fruit surfaces of different apple cultivars. *Arthropod Plant Interact.* **2014**, *8*, 57–77. [CrossRef]

46. Voigt, D.; Gorb, S.N. Egg attachment of the asparagus beetle *Crioceris asparagi* to the crystalline waxy surface of *Asparagus officinalis*. *Proc. R. Soc. B* **2010**, *277*, 895–903. [CrossRef] [PubMed]

47. Fordyce, J.A.; Nice, C.C. Variation in butterfly egg adhesion: Adaptation to local host plant senescence characteristics? *Ecol. Lett.* **2003**, *6*, 23–27. [CrossRef]

48. Li, D.; Huson, M.G.; Graham, L.D. Proteinaceous adhesive secretions from insects; in particular the egg attachment glue of *Opodiphthera* sp. moths. *Arch. Insect Biochem.* **2008**, *69*, 85–105. [CrossRef]

49. Yoshida, K.; Nagata, M. Adhesive strength of the glue substances in the colleterial glands of the silkmoth, *Bombyx mori*. *J. Seric. Sci. Jpn.* **1997**, *66*, 453–456. [CrossRef]

50. Yago, M.; Mitamura, T.; Abe, S.; Hashimoto, S. Adhesive strength of glue-like substances from the colleterial glands of *Antheraea yamamai* and *Rhodinia fugax*. *Int. J. Wild Silkmoth Silk* **2001**, *6*, 11–15.

51. England, M.W.; Sato, T.; Yagihashi, M.; Hozumi, A.; Gorb, S.N.; Gorb, E.V. Surface roughness rather than surface chemistry essentially affects insect adhesion. *Beilstein J. Nanotechnol.* **2016**, *7*, 1471–1479. [CrossRef]

52. Habenicht, G. *Kleben: Grundlagen, Technologien, Anwendung*, 4th ed.; Springer: Berlin, Germany, 2002.

53. Cogley, T.P.; Anderson, J.R.; Weintraub, J. Ultrastructure and function of the attachment organ of warble fly eggs (Diptera: Oestridae: Hypodermatinae). *Int. J. Insect Morphol. Embryol.* **1981**, *10*, 7–18. [CrossRef]

54. Cogley, T.P.; Cogley, M.C. Morphology of the eggs of the human bot fly, *Dermatobia hominis* (L. Jr.) (Diptera: Cuterebridae) and their adherence to the transport carrier. *Int. J. Insect Morph. Embryol.* **1989**, *18*, 239–248. [CrossRef]

55. Hinton, H.E. *Biology of Insect Eggs*; Pergamon Press: Oxford, UK, 1981. [CrossRef]

56. Margaritis, L.H. Structure and physiology of eggshell. In *Comprehensive Insect Physiology, Biochemistry and Pharmacology*; Kerkut, G.A., Gilbert, L.I., Eds.; Pergamon Press: Oxford, UK, 1985; pp. 153–230.

57. Miller, P.L. Oviposition behaviour and eggshell structure in some libellulid dragonflies, with particular reference to *Brachythemis lacustris* (Kirby) and *Orthetrum coerulescens* (Fabricius)(Anisoptera). *Odonatologica* **1987**, *16*, 361–374.

58. Ivey, R.K.; Bailey, J.C.; Stark, B.P.; Lentz, D.L. A preliminary report of egg chorion features in dragonflies (Anisoptera). *Odonatologica* **1988**, *17*, 393–399.

59. Trueman, J.W.H. Egg chorionic structures in Corduliidae and Libellulidae (Anisoptera). *Odonatologica* **1991**, *20*, 441–452.

60. Sahlén, G. Ultrastructure of the eggshell and micropylar apparatus in *Somatochlora metallica* (Vander, L.), *Orthetrum cancellatum* (L.) and *Sympetrum sanguineum* (Müll.) (Anisoptera: Corduliidae, Libellulidae). *Odonatologica* **1994**, *23*, 255–269.

61. Sahlén, G. Eggshell ultrastructure in *Onychogomphus forcipatus unguiculatus* (Vander Linden) (Odonata: Gomphidae). *Int. J. Insect Morphol. Embryol.* **1995**, *24*, 281–286. [CrossRef]

62. Andrew, R.J.; Tembhare, D.B. Ultrastructural post-oviposition changes in the egg chorion of the dragon-fly, *Zyxomma petiolatum* Rambur (Odonata: Libellulidae). *Int. J. Insect Morphol. Embryol.* **1995**, *24*, 235–238. [CrossRef]

63. Andrew, R.J.; Tembhare, D.B. Surface ultrastructure of the egg chorion of *Bradinopyga geminata* (Rambur) and *Rhyothemis variegata variegata* (Linn.). *Fraseria* **1996**, *3*, 1–5.

64. Andrew, R.J. Egg chorionic ultrastructure of the dragonfly *Tramea virginia* (Rambur) (Anisoptera: Libellulidae). *Odonatologica* **2002**, *31*, 171–175.

65. Gaino, E.; Piersanti, S.; Rebora, M. Egg envelope synthesis and chorion modification after oviposition in the dragonfly *Libellula depressa* (Odonata, Libellulidae). *Tissue Cell* **2008**, *40*, 317–324. [CrossRef]

66. Gaino, E.; Rebora, M. Synthesis and function of the fibrous layers covering the eggs of *Siphlonurus lacustris* (Ephemeroptera, Siphlonuridae). *Acta Zool.* **2001**, *82*, 41–48. [CrossRef]

67. Brinck, P. Reproductive system and mating in Ephemeroptera. *Opusc. Entomol.* **1957**, *22*, 1–37.

68. Koss, R.W. Ephemeroptera Eggs: Sperm Guide Morphology and Adhesive Layer Formation. *Trans. Am. Micros. Soc.* **1970**, *89*, 295–299. [CrossRef]

69. Gaino, E.; Mazzini, M. Scanning electron microscopy of the egg attachment structures of *Electrogena zebrata* (Ephemeroptera: Heptageniidae). *Trans. Amer. Micros. Soc.* **1987**, *106*, 114–119. [CrossRef]

70. Gaino, E.; Mazzini, M. Fine structure of the chorionic projections of the egg of *Rhithrogena kimminsi* Thomas (Ephemeroptera: Heptageniidae) and their role in egg adhesion. *Int. J. Insect Morphol. Embryol.* **1988**, *17*, 113–120. [CrossRef]

71. Gaino, E.; Mazzini, M. Chorionic adhesive material of the egg of the mayfly *Habrophlebia eldae* (Ephemeroptera, Leptophlebiidae): Morphology and synthesis. *Boll. Zool.* **1989**, *56*, 291–298. [CrossRef]

72. Wohlfart, E.; Wolff, J.O.; Arzt, E.; Gorb, S.N. The whole is more than the sum of all its parts: Collective effect of spider attachment organs. *J. Exp. Biol.* **2014**, *217*, 222–224. [CrossRef]

73. Spurr, A.R. A low-viscosity epoxy resin embedding medium for electron microscopy. *J. Ultrastruct. Res.* **1969**, *26*, 31–43. [CrossRef]

74. Salerno, G.; Rebora, M.; Gorb, E.V.; Kovalev, A.E.; Gorb, S.N. Attachment ability of the southern green stink bug *Nezara viridula* (Heteroptera: Pentatomidae). *J. Comp. Physiol. A* **2017**, *203*, 601–611. [CrossRef]

75. Schaber, C.F.; Kreitschitz, A.; Gorb, S.N. Friction-active surfaces based on free-standing anchored cellulose nanofibrils. *ACS Appl. Mater. Interfaces* **2018**, *10*, 37566–37574. [CrossRef]

76. Santos, R.; Gorb, S.N.; Jamar, V.; Flammang, P. Adhesion of echinoderm tube feet to rough surfaces. *J. Exp. Biol.* **2005**, *208*, 2555–2567. [CrossRef] [PubMed]

77. Gorb, S.N. Biological attachment devices: Exploring nature's diversity for biomimetics. *Phil. Trans. Math. Phys. Eng. Sci.* **2008**, *366*, 1557–1574. [CrossRef]

78. Scherge, M.; Gorb, S.N. *Biological Micro- and Nanotribology*; Springer: Heidelberg/Berlin, Germany, 2001.

79. Filippov, A.; Popov, V.L.; Gorb, S.N. Shear induced adhesion: Contact mechanics of biological spatula-like attachment devices. *J. Theor. Biol.* **2011**, *276*, 126–131. [CrossRef]

80. Persson, B.N.J. On the mechanism of adhesion in biological systems. *J. Chem. Phys.* **2003**, *118*, 7614–7621. [CrossRef]

81. Persson, B.N.J.; Gorb, S.N. The effect of surface roughness on the adhesion of elastic plates with application to biological systems. *J. Chem. Phys.* **2003**, *119*, 11437–11444. [CrossRef]

82. Johnson, K.L.; Kendall, K.; Roberts, A.D.; Tabor, D. Surface energy and the contact of elastic solids. *Proc. R. Soc. A* **1971**, *324*, 301–313. [CrossRef]

83. Arzt, E.V.; Gorb, S.N.; Spolenak, R. From micro to nano contacts in biological attachment devices. *Proc. Natl. Acad. Sci. USA* **2003**, *100*, 10603–10606. [CrossRef]

84. Gorb, S.N.; Heepe, L. Biological fibrillar adhesives: Functional principles and biomimetic applications. In *Handbook of Adhesion Technology*; da Silva, L., Öchsner, A., Adams, R., Eds.; Springer: Cham, Switzerland, 2017; pp. 1–37.

85. Gorb, S.N. *Attachment Devices of Insect Cuticle*; Kluwer Academic: Dordrecht, The Netherlands, 2001.

86. Peressadko, A.; Gorb, S.N. Surface profile and friction force generated by insects. In Proceedings of the First International Industrial Conference Bionik 2004, Hannover, Germany, 22–23 April 2004; Boblan, I., Bannasch, R., Eds.; VDI Verlag: Düsseldorf, Germany, 2004; pp. 257–263.

87. Voigt, D.; Schuppert, J.M.; Dattinger, S.; Gorb, S.N. Sexual dimorphism in the attachment ability of the Colorado potato beetle *Leptinotarsa decemlineata* (Coleoptera: Chrysomelidae) to rough substrates. *J. Insect Physiol.* **2008**, *54*, 765–776. [CrossRef]

88. Bullock, J.M.R.; Federle, W. Division of labour and sex differences between fibrillar, tarsal adhesive pads in beetles: Effective elastic modulus and attachment performance. *J. Exp. Biol.* **2009**, *212*, 1876–1888. [CrossRef]

89. Gorb, E.V.; Gorb, S.N. Effects of surface topography and chemistry of *Rumex obtusifolius* leaves on the attachment of the beetle *Gastrophysa viridula*. *Entomol. Exp. Appl.* **2009**, *130*, 222–228. [CrossRef]

90. Gorb, E.V.; Hosoda, N.; Miksch, C.; Gorb, S.N. Slippery pores: Anti-adhesive effect of nanoporous substrates on the beetle attachment system. *J. R. Soc. Interface* **2010**, *7*, 1571–1579. [CrossRef] [PubMed]

91. Prüm, B.; Seidel, R.; Bohn, H.F.; Speck, T. Plant surfaces with cuticular folds are slippery for beetles. *J. R. Soc. Interface* **2012**, *9*, 127–135. [CrossRef] [PubMed]

92. Ramaswamy, S.B.; Ma, W.K.; Baker, G.T. Sensory cues and receptors for oviposition by *Heliothis virescens*. *Entomol. Exp. Appl.* **1987**, *43*, 159–168. [CrossRef]

93. Ramaswamy, S.B. Host finding by moths: Sensory modalities and behaviours. *J. Insect Physiol.* **1988**, *34*, 235–249. [CrossRef]

94. Hilker, M.; Meiners, T. Early herbivore alert: Insect eggs induce plant defense. *J. Chem. Ecol.* **2006**, *32*, 1379–1397. [CrossRef]

95. Grohmann, C.; Blankenstein, A.; Koops, S.; Gorb, S.N. Attachment of *Galerucella nymphaeae* (Coleoptera, Chrysomelidae) to surfaces with different surface energy. *J. Exp. Biol.* **2014**, *217*, 4213–4220. [CrossRef]

96. Israelachvili, J. *Intermolecular and Surface Forces*; Academic Press: London, UK, 1992; pp. 415–467.

97. Kendall, K. The adhesion and surface energy of elastic solids. *J. Phys. D* **1971**, *4*, 1186. [CrossRef]

98. Lüken, D.; Voigt, D.; Gorb, S.N.; Zebitz, C.P.W. Die Tarsenmorphologie und die Haftfähigkeit des Schwarzen Batatenkäfers *Cylas puncticollis* (Boheman) auf glatten Oberflächen mit unterschiedlichen physiko-chemischen Eigenschaften. *Mitt. Dtsch. Ges. Allg. Angew. Ent.* **2009**, *17*, 109–113.

99. Voigt, D.; Gorb, S.N. Attachment ability of sawfly larvae to smooth surfaces. *Arthropod Struct. Dev.* **2012**, *41*, 145–153. [CrossRef]

100. Thomas, J.; Peppas, N. Adhesives. In *Encyclopedia of Biomaterials and Biomedical Engineering*; Wnek, G.E., Bowlin, G.L., Eds.; Taylor & Francis: Abingdon, UK; CRC Press: Boca Raton, FL, USA, 2008; pp. 1–7. [CrossRef]

101. Woolman, C.; Dhannasiri, B. Food plants for *Phyllium bioculatum* Gray in Sri Lanka. *Phasmid Stud.* **1995**, *4*, 33.

102. Wang, H.; Shi, H.; Li, Y.; Yu, Y.; Zhang, J. Seasonal variations in leaf capturing of particulate matter, surface wettability and micromorphology in urban tree species. *Front. Env. Sci. Eng.* **2013**, *7*, 579–588. [CrossRef]

103. Mohammed-Ziegler, I.; Oszlánczi, Á.; Somfai, B.; Hórvölgyi, Z.; Pászli, I.; Holmgren, A.; Forsling, W. Surface free energy of natural and surface-modified tropical and European wood species. *J. Adhes. Sci. Technol.* **2004**, *18*, 687–713. [CrossRef]

104. Prüm, B.; Bohn, H.F.; Seidel, R.; Rubach, S.; Speck, T. Plant surfaces with cuticular folds and their replicas: Influence of microstructuring and surface chemistry on the attachment of a leaf beetle. *Acta Biomater.* **2013**, *9*, 6360–6368. [CrossRef] [PubMed]

105. Malhi, Y.; Wright, J. Spatial patterns and recent trends in the climate of tropical rainforest regions. *Philos. Trans. R. Soc. B* **2004**, *359*, 311–329. [CrossRef] [PubMed]

106. Arai, M.; Yago, M. Curious oviposition behavior in *Phyllium westwoodii* (Phasmatodea: Phylliidae): Preliminary observations. *J. Insect Sci.* **2015**, *15*, 135. [CrossRef] [PubMed]

107. Rosen, M.J.; Kunjappu, J.T. *Surfactants and Interfacial Phenomena*; Wiley: Hoboken, NJ, USA, 2012.

108. Subramani, K.; Ahmed, W. Self-assembly of proteins and peptides and their applications in bionanotechnology and dentistry. In *Emerging Nanotechnologies in Dentistry 2*; Elsevier: Amsterdam, The Netherlands, 2018; pp. 231–249. [CrossRef]

109. Beament, J.W.L.; Lal, R. Penetration through the Egg-shell of *Pieris brassicae* (L.). *Bull. Entomol. Res.* **1957**, *48*, 109–125. [CrossRef]

110. Riley, R.C.; Forgash, A.J. *Drosophila melanogaster* eggshell adhesive. *J. Insect Physiol.* **1967**, *13*, 509–517. [CrossRef]

111. Amornsak, W.; Noda, T.; Yamashita, O. Accumulation of glue proteins in the developing colleterial glands of the silkworm, *Bombyx mori*. *J. Seric. Sci. Jpn.* **1992**, *61*, 123–130. [CrossRef]

112. Burkhart, C.N.; Stankiewicz, B.A.; Pchalek, I.; Kruge, M.A.; Burkhart, C.G. Molecular composition of the louse sheath. *J. Parasitol.* **1999**, *85*, 559–561. [CrossRef]

113. Jin, Y.; Chen, Y.; Jiang, Y.; Xu, M. Proteome analysis of the silkworm (*Bombyx mori*. L) colleterial gland during different development stages. *Arch. Insect Biochem. Physiol.* **2006**, *61*, 42–50. [CrossRef]

114. Betz, O. Adhesive Exocrine Glands in Insects: Morphology, Ultrastructure; Adhesive Secretion. In *Biological Adhesive, Systems*; von Byern, J., Grunwald., I., Eds.; Springer: Vienna, Austria, 2010; pp. 111–152. [CrossRef]

115. Burgess, I.F. Do nit removal formulations and other treatments loosen head louse eggs and nits from hair? *Med. Vet. Entomol.* **2010**, *24*, 55–61. [CrossRef] [PubMed]

116. Graham, L.D. Biological adhesives from nature. In *Encyclopedia of Biomaterials and Biomedical Engineering*; Wnek, G.E., Bowlin, G.L., Eds.; Taylor & Francis: Abingdon, UK; CRC Press: Boca Raton, FL, USA, 2008; pp. 236–253.

117. Dalziel, M.; Crispin, M.; Scanlan, C.N.; Zitzmann, N.; Dwek, R.A. Emerging principles for the therapeutic exploitation of glycosylation. *Science* **2014**, *343*, 1235681. [CrossRef] [PubMed]

118. Xi, E.; Venkateshwaran, V.; Li, L.; Rego, N.; Patel, A.J.; Garde, S. Hydrophobicity of proteins and nanostructured solutes is governed by topographical and chemical context. *Proc. Natl. Acad. Sci. USA* **2017**, *114*, 13345–13350. [CrossRef]

119. Beňová-Liszeková, D.; Beňo, M.; Farkaš, R. Fine infrastructure of released and solidified *Drosophila* larval salivary secretory glue using SEM. *Bioinspir. Biomim.* **2019**, *14*, 055002. [CrossRef]

120. Borne, F.; Kovalev, A.E.; Gorb, S.N.; Courtier-Orgogozo. The glue produced by *Drosophila melanogaster* for pupa adhesion is universal. *J. Exp. Biol.* **2020**, *223*, jeb220608. [CrossRef] [PubMed]

121. Vandermeer, J.H.; Goldberg, D.E. *Population Ecology: First Principles*, 2nd ed.; Princeton University Press: Princeton, NJ, USA, 2013.

122. Rentz, D.C. *Grasshopper Country: The Abundant Orthopteroid Insects of Australia*; UNSW Press: Sydney, Australia, 1996.

123. Patel, A.K.; Mathias, J.-D.; Michaud, P. Polysaccharides as adhesives. *Rev. Adhes. Adhes.* **2013**, *1*, 312–345. [CrossRef]

124. Karak, N. Biopolymers for paints and surface coatings. In *Biopolymers and Biotech Admixtures for Eco-Efficient Construction Materials*; Pacheco-Torgal, F., Ivanov, V., Karak, N., Jonkers, H., Eds.; Woodhead Publishing: Cambridge, UK, 2016; pp. 333–368. [CrossRef]

 *insects*

Article

The Effect of Ground Type on the Jump Performance of Adults of the Locust *Locusta migratoria manilensis*: A Preliminary Study

Chao Wan [1,2,*], Rentian Cao [1] and Zhixiu Hao [1,*]

[1] State Key Laboratory of Tribology, Tsinghua University, Beijing 100084, China
[2] Department of Mechanics, School of Aerospace Engineering, Beijing Institute of Technology, Beijing 100081, China
*   Correspondence: chaowan@bit.edu.cn (C.W.); haozx@tsinghua.edu.cn (Z.H.)

Received: 18 March 2020; Accepted: 12 April 2020; Published: 23 April 2020

Abstract: The jump performance of locusts depends on several physiological and environmental factors. Few studies have examined the effects of different ground types on the jump performance of locusts. Here, mature adult locusts (*Locusta migratoria manilensis*) were examined using a custom-developed measuring system to test their jump performance (including postural features, kinematics, and reaction forces) on three types of ground (sand, soil, and wood). Significant differences were primarily observed in the elevation angle at take-off, the tibial angle at take-off, and the component of the mass-specific reaction force along the aft direction of the insect body between wood and the other two ground types (sand and soil). Slippage of the tarsus and insertion of the tibia were often observed when the locusts jumped on sand and soil, respectively. Nevertheless, comparisons of the different parameters of jump initiation (i.e., take-off speed and mass-specific kinetic energy) did not reveal any differences among the three types of ground, indicating that locusts were able to achieve robust jump performance on various substrates. This study provides insights into the biomechanical basis of the locust jump on different types of ground and enhances our understanding of the mechanism underlying the locust jump.

Keywords: ground type; jump; locust; reaction force; kinematics; elevation angle

1. Introduction

Locusts are some of the most famous insects for their jumping ability, as they can achieve velocities as high as 2.6 m/s, accelerations as high as 75 m/s^2, and can cover dozens of times their body length in a single jump [1]. Their jumps serve several critical functions: to escape from predators, to achieve an initial velocity for flight, and provide a more rapid alternative to travel than crawling. The locust jump has been extensively studied, especially its postural control, the mechanics of the hind leg, patterns of muscle and motoneuron activity, and mechanisms of energy storage and release [2–11]. The action of jumping in locusts is fueled by their hind legs in the following steps: initial flexion of the tibiae, co-contraction of the flexor and extensor muscles, and rapid tibial extension after trigger activity [2]. A large amount of strain energy is stored by the deformed exoskeleton during muscle co-contraction (especially by the semi-lunar process (SLP) cuticle on the distal end of the metathoracic femur) and is released during tibial extension to overcome the weaknesses of insect muscles [3,4].

Various physiological factors have been shown to have different effects on the jumping performance of locusts. Compared with immature juvenile hoppers, adult locusts have three times the range of escape jumps and at least twice the specific energy output [12,13]. Older juvenile insects hop less frequently than younger ones within the same instar because of increased anaerobic metabolism and locomotory fatigue [14]. Adaptive change in muscle contraction has also been observed in newly molted locusts to avoid cuticle damage during jumping [5]. In contrast, gravid females (20% heavier) have the same jump distance and significantly lower endurance compared with non-gravid females, as non-gravid females show a 20% increase in the duration of muscle contraction relative to gravid females [15,16]. Furthermore, Katz and Gosline [17] found that the take-off speed of the locust jump is relatively scale-independent (0.9–1.2 m/s for juveniles and 2.5 m/s for adults), showing that juvenile insects often jump to maximize the distance traveled while the purpose of jumping in adults more often serves to achieve a velocity necessary for initiating flight.

In addition, the effects of a few natural environmental factors have also been investigated. Hawlena et al. [18] reported that the chronic risk of predation can increase both the take-off speed and the jump distance of grasshoppers and that this pattern cannot be explained by morphological variation. Air resistance reduces the kinetic energy of locust jumps by less than 10% at lower initial speeds [19], and environmental temperatures ranging from 15 °C to 35 °C only weakly affect jump energy [20]. The properties of the ground are another set of environmental factors that can potentially affect the locust jump, especially during the take-off stage. For example, surface roughness with an Ra value of 1–2 μm can reduce the ability of locust legs to attach to the substrate, resulting in considerable slippage of the hind legs on the ground and thus take-off failure [21,22]. Similar effects of surface roughness have also been documented in females of the Mediterranean field cricket (*Gryllus bimaculatus*) crawling on smooth surfaces (R_q = 7.3 μm), which resulted in significantly lower phonotactic responses compared with rougher surfaces (R_q = 16 or 180 μm) [23]. Several natural types of ground surfaces have other key physical properties as well as surface roughness, such as normal stiffness, hardness, and tangential friction/shear stress strength. However, whether natural ground types can affect the jump performance of locusts has not been explored.

Here, we compared the kinematics, insect posture, and reaction force of the jump performance of adult locusts on three natural ground types (sand, soil, and wood). We hypothesized that the jump performance of locusts would differ among the three types of ground. Our results provide insight into the biomechanical basis of the locust jump on different types of ground and enhance our understanding of the mechanisms underlying the locust jump.

2. Materials and Methods

Mature Oriental migratory locusts (*Locusta migratoria manilensis*) were purchased from the Jiyuan locust-breeding facility in Anhui Province of China. Before tests, all locusts were fed with wheat leaves under natural light and room temperature (23–30 °C) for at least 2 weeks to ensure fully sclerotized cuticles after their final molt.

A custom-developed test system was built to simultaneously measure both the kinematics and reaction force of the locust jumps. Figure 1 shows the components of the test system, including a coordinate background plate (indicated by the letter G), two high-speed cameras (CR600 × 2, Optronis GmbH, Kehl, Germany; Hero4, GoPro Inc., San Mateo, CA, USA; indicated by letters D and E), a video camera (IXUS 210, Canon Inc., Tokyo, Japan; indicated by the letter F), a desktop for data acquisition (indicated by the letter H), and two custom-designed platforms (indicated by the letter B) supported by a stand column and a three-dimensional high-precision force sensor (S3-001NTO-003, Bio-inspired Technology Corp., Nanjing, China; indicated by the letter A). The two platforms were set in the same horizontal plane with 1 mm of space. The two high-speed cameras were placed on the side of the platforms. The GoPro camera (120 fps) was used to obtain the jump trajectory of the animals, and the Optronis camera (1000 fps) was used to record the rapid movement of their hind legs before take-off. The Canon camera was placed above the system to take images of insect posture and the jump azimuth. The force sensor was used to measure the three-dimensional reaction force from one hind leg of the locust during jumping (full-scaled range of 1 N, resolution of 1 mN, and sample frequency of 10 kHz).

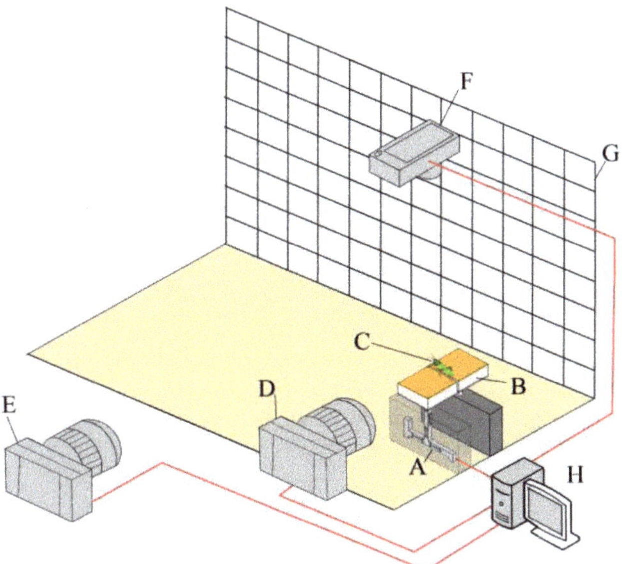

Figure 1. A custom-developed test system for measuring the kinematics and reaction force of adult locusts while jumping. (**A**) Three-dimensional high-precision force sensor; (**B**) two platforms, one of which is fixed to the force sensor and the other fixed to a stand column; (**C**) adult locust whose left and right hind legs each stood on one platform; (**D**) a high-speed camera for obtaining the rapid movement of the hind leg during the jump; (**E**) a high-speed camera for obtaining the trajectory of the locust jump; (**F**) a video camera for imaging the posture of the locust from above; (**G**) a coordinate background plate; and (**H**) desktop computer for data acquisition.

Three ground types (sand, soil, and wood) were studied. First, balsa wood was fixed on the surface of the platforms to simulate wooden ground. The balsa wood was cut from some timbers and polished using moderate sandpaper (#: P400). All obvious visual burrs were removed from the wood surface. The locust was prepared by attaching its wings together using adhesive tape and placing it on the platforms such that its left and right hind legs each stood on one platform. Either a sudden sound or the touch of a brush was used to trigger the locust's escaping jump. Data were excluded for individuals that had two hind legs on the platform of the force sensor. Next, the balsa wood was replaced by two boxes (width × length × depth = 15 mm × 40 mm × 10 mm) that were filled with local sand (grain size of approximately 0.4 mm, density = 1.27 g/cm^3) to create a sandy substrate or commercial potting soil (soft and rich in organic matter, density = 0.317 g/cm^3) to create a soil substrate. Both of the fillings were lightly compacted, and the surface was carefully flattened before tests. Jumps of the locusts on the sand and the soil were tested following the aforementioned procedures. In total, nine adult locusts (4 females and 5 males) were included (body mass = 1.8 ± 0.51 g (mean ± SD)) and used for all the three ground types in this study. For avoiding the animal fatigue, only one ground type was tested in one day. For each ground type, each animal was tested less than seven trials and its first successful jump was selected to represent the jump performance of this individual. In brief, nine jumps were included for each ground type (one jump from each animal for each ground type). The sand, soil, and wood ground types corresponded to a granular substrate with weak cohesion, a granular substrate with strong cohesion, and a solid substrate, respectively.

The yaw angle of a jump (φ) was obtained from the top camera image as the angle between the body axis of the locust and the direction X of the force sensor (Figure 2). The real distance of each locust jump (S_d) was then obtained by correcting the camera recordings using the following equation:

$$S_d = \frac{\overline{S}_d}{\cos(\varphi)} \tag{1}$$

where $\overline{S}_d$ is the distance recordings from the lateral images. The physiological components of the reaction force for the locusts were calculated from the force measurements of the sensor using the following equation:

$$\overline{F}_a = F_x \cos(\varphi) - F_y \sin(\varphi), \overline{F}_l = F_x \sin(\varphi) + F_y \cos(\varphi), \overline{F}_n = F_z \tag{2}$$

where F_x, F_y, and F_z are the measurements from the force sensor along its axes X, Y, and Z, respectively. $\overline{F}_a$, $\overline{F}_l$, and $\overline{F}_n$ are the physiological components of the reaction force along the aft, lateral, and normal directions of the locust, respectively (Figure 2b). To eliminate the effect of body mass, the mass-specific reaction force was calculated by normalizing the physiological components with body mass as follows:

$$F_a = \frac{\overline{F}_a}{M}, F_l = \frac{|\overline{F}_l|}{M}, F_n = \frac{\overline{F}_n}{M}, F_t = \sqrt{F_a^2 + F_l^2 + F_n^2} \tag{3}$$

where M is body mass. F_a, F_l, and F_n are the aft, lateral, and normal components of the mass-specific reaction force, respectively. F_t is the total magnitude of the mass-specific reaction force. The elevation angle of the locust at take-off (β_t) was determined based on the real trajectory of the jump from the images of the lateral GoPro camera after take-off (Figure 2).

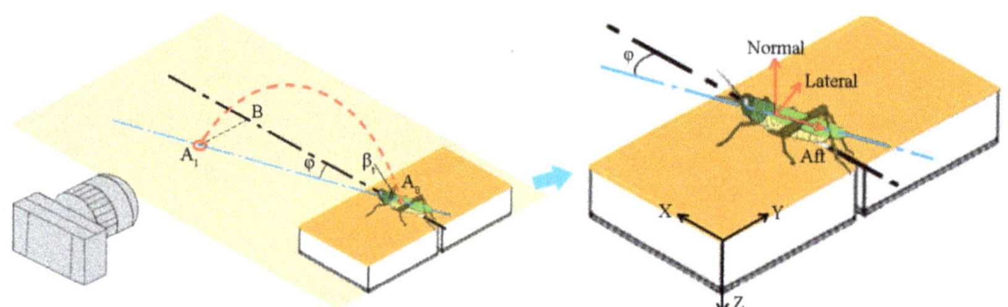

Figure 2. Schematic diagram illustrating how the kinematics and reaction force of the locust jumps were calculated. Because of the yaw angle (φ), the real trajectory of the jump was corrected from the image recordings of the sideway high-speed camera. The jump direction and real trajectory of the locust are indicated by the blue and red dashed lines, respectively. The black dashed line is parallel to both the lateral camera and the direction X of the force sensor. A_0: the initial locust position; A_1: the real locust position after jumping; B: the locust position in the camera images after jumping; and β_t: the elevation angle of the locust at take-off. Similarly, the physiological components of the reaction force were transformed from the sensor measurements using the yaw angle. The coordinate system of the force sensor is shown by black axes, and the physiological coordinate system of the locust is shown by red axes.

Three postural features of the locust during jumps were defined based on the camera images as follows (Figure 3). First, the opening angle between the hind femur and the central line of the body (θ) was measured from the image of the top camera just before tibial extension. Following a previous study [6], a line was drawn through the proximal femoral and the distal tibial ends of the locust just before tibial extension, and its tilt angle relative to the ground surface was defined as β_e. Here, the angle β_e was determined by both the projected tilt angle in the Optronis camera image ($\overline{\beta}_e$) and the yaw angle φ as $\arctan(\cos(\varphi)\tan(\overline{\beta}_e))$. Similarly, the angle between the tibial axis and the ground surface at take-off (γ) was calculated as $\arctan(\cos(\varphi)\tan(\overline{\gamma}))$, where $\overline{\gamma}$ is the projected tibia–ground angle in the Optronis camera image. Given the small waste in energy because of air resistance [19], the take-off speed (V_t) was calculated based on the real jump distance S_d and the angle β_t as

$$V_t = \sqrt{\frac{gS_d\tan(\beta_t)}{2} + \frac{gS_d}{2\tan(\beta_t)}} \tag{4}$$

where g is the acceleration of gravity (i.e., 9.8 m s^{-2}). The mass-specific kinetic energy (E_m) for each jump was determined as $0.5V_t^2$.

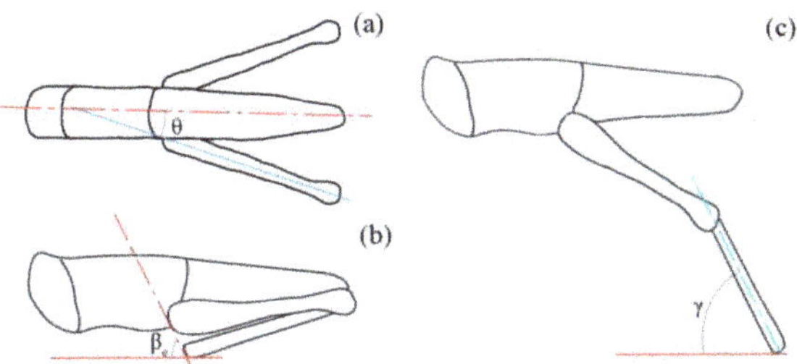

Figure 3. Schematic diagram for three postural features of adult locusts while jumping. (**a**) The opening angle between the hind femur and central line of the body just before tibial extension (θ); (**b**) the tilt angle of the line through the proximal femur and distal tibia just before tibial extension (β_e); and (**c**) the tibial angle relative to the ground at take-off (γ).

Statistical analyses were performed to clarify the effect of ground type on the locust jump. First, the normal distribution and homoscedasticity of the results were checked using Shapiro–Wilk tests and F-tests/Bartlett's tests, respectively. If the data met these requirements, t-tests were performed for all of the data between female and male animals and between the angles β_t and β_e for each ground type. Repeated one-way analyses of variance (ANOVA) with a Bonferroni post-hoc correction were also used to compare results between the three different ground types. Mann–Whitney tests were used instead of t-tests if the data were not normally distributed, and t-tests with Welch's correction were used if the data had unequal variances. For repeated one-way ANOVA, Friedman tests with Dunn's multiple comparisons were used if the data were either not normally distributed or variances were not homogeneous. Results were reported as mean ± SD. Significance was defined as $p < 0.05$.

3. Results

Both the postural features (θ, γ, β_t, and β_e) and jump performance (V_t and E_m), including the reaction force (the maximal values of F_a, F_l, F_n, and F_t), of adult female locusts were not significantly ($p > 0.05$) different from those of adult male locusts regardless of ground type. As a result, data for both female and male locusts were pooled and analyzed together in subsequent statistical analyses. All of the experimental data are shown in Tables S1–S6 of the Supplementary Materials.

The four postural angles of the locust jumps on the three ground types are shown in Figure 4. The angles β_e and β_t were 57 ± 7.3 degrees and 36 ± 9.5 degrees, 52 ± 9.2 degrees and 43 ± 6.2 degrees, and 64 ± 12 degrees and 63 ± 12 degrees for the sand, soil, and wood substrates, respectively. Significant differences were detected in the angle β_e between soil and wood ($p < 0.05$), in the angle β_t between sand and wood ($p < 0.001$), and in the angle β_t between soil and wood ($p < 0.01$). Comparisons between these two angles indicated that differences were only significant between sand and soil ($p < 0.001$). Moreover, the angle γ for the wood (68 ± 10 degrees) was significantly ($p < 0.001$) higher than that for the sand (44 ± 8.6 degrees) as well as that for the soil (51 ± 6.5 degrees). By contrast, the angle θ was not significantly affected by changes in ground type.

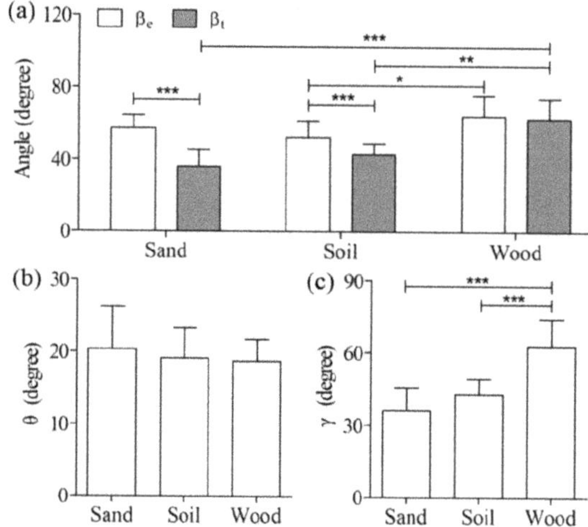

Figure 4. Comparisons of the postural features of adult locust jumps on the three ground types: (**a**) the elevation angle of locust jumps at take-off (β_t) and the tilt angle of the line through femoral proximal and tibial distal ends just before tibial extension (β_e); (**b**) the open angle of the femur to the central line of the body (θ); and (**c**) the angle between the tibial axis and the ground at take-off (γ). Results were reported as mean ± SD. * $p < 0.05$; ** $p < 0.01$; *** $p < 0.001$.

Figure 5 shows comparisons of the kinematics and reaction force among the three ground types. Neither the take-off speed (V_t) nor the mass-specific kinetic energy (E_m) was significantly different among the three ground types (2.3 ± 0.33 m/s and 2.7 ± 0.83 mJ/g for sand, 2.3 ± 0.13 m/s and 2.7 ± 0.31 mJ/g for soil, and 2.6 ± 0.17 m/s and 3.3 ± 0.42 mJ/g for wood). There were no significant differences in the mass-specific reaction force among the three ground types except for F_a. The F_a for wood (20 ± 6.3 mN/g) was significantly ($p < 0.05$) lower than that for sand (38 ± 5.6 mN/g) and soil (31 ± 10 mN/g). In addition, the hind legs of the locusts during the jumps firmly adhered to the wood substrate, often inserted into the sandy substrate (six of the nine jumps), or noticeably slid on the soil substrate (seven of the nine jumps). To illustrate the different interactions between the hind legs of locusts and the three ground types, some typical high-speed camera images and videos of the locust hind legs during jumping are provided in Figure 6 and the Supplementary Videos S1–S3.

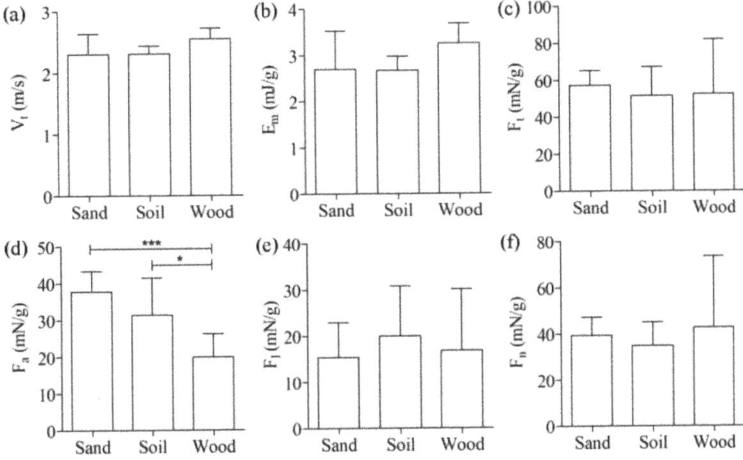

Figure 5. Comparisons of the kinematics and reaction force of locust jumps on three ground types: (**a**) the take-off speed (V_t); (**b**) the mass-specific kinetic energy (E_m); (**c**) the total magnitude of the mass-specific reaction force (F_t); (**d**) the aft component of the mass-specific reaction force (F_a); (**e**) the absolute lateral component of the mass-specific reaction force (F_l); and (**f**) the normal component of the mass-specific reaction force (F_n). Results were reported as mean ± SD. * $p < 0.05$; *** $p < 0.001$.

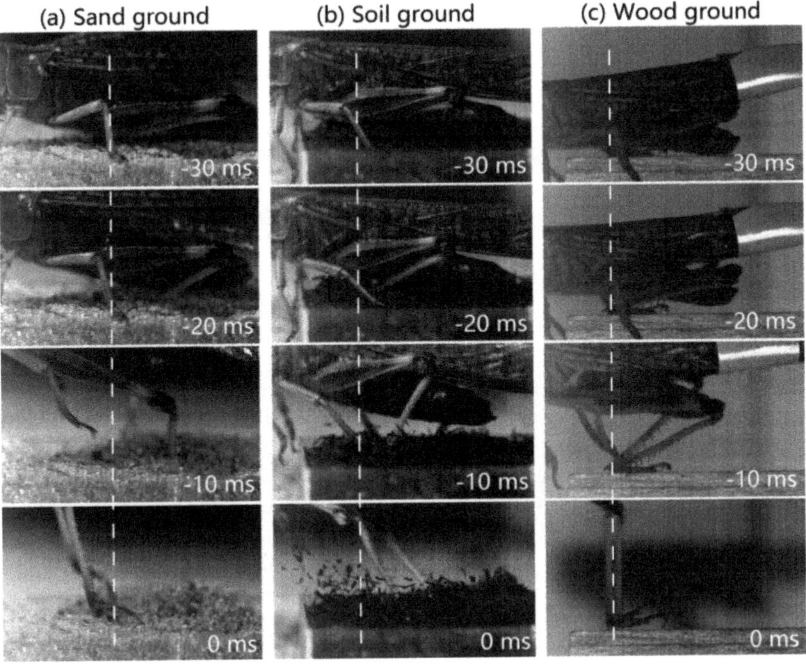

Figure 6. High-speed camera images for the rapid movement of the hind legs of locusts while jumping. Locusts jumped on sand (**a**), soil (**b**), and wood (**c**) ground types. The frame at 0 ms shows the configuration of the hind leg at take-off. To analyze the jump process, white dashed lines were marked for each panel to represent the location of the tibial distal end at 0 ms in the view of the camera.

4. Discussion

We suggest that locusts have a robust jump sequence on various substrates. Both the take-off speed and the mass-specific kinetic energy of the locust jump were not significantly affected by the change in ground type even though tibial slip or insertion occurred on sand and soil substrates during the initial stage of jumping (Figure 6b,c). Given that the main purpose of jumping for adult locusts is to achieve a sufficient initial speed for initiating flight [17], the fact that the take-off speed did not change among the different ground types revealed that the jump behaviors of the locust were robust on various substrates. The kinetic energy of the locust jump was primarily released from the elastic strain energy stored by the SLP cuticle during the extension of the hind leg [3,4]. The similar mass-specific kinetic energy for the jumps on the three ground types meant that the stored strain energy was not wasted by the somewhat useless extension arising from the slip or insertion of tibia. This finding coincides with the previous observation that the semi-lunar processes of locusts do not start to unfurl (i.e., release energy) until the tibia extends by 55 degrees [24]. In other words, the specific mechanism underlying the storage and release of strain energy helps locusts achieve high jumping performance on different ground types because the energy does not begin to release during the initial extension, regardless as to whether the tibia slips on or inserts into the substrate. This property might enhance the ability of locusts to evade predators and thus their survival in different environments.

By contrast, the postural features of locusts among the three types of ground indicated that significant differences were primarily observed in the elevation angle at take-off (β_t) and the tibial angle at take-off (γ) (Figure 4). Sutton and Burrows [6] found that the angle β_t was almost the same as the tilt angle of the femoral proximal femur–distal tibia line before tibial extension (β_e) when locusts jumped on wood. Similar results were obtained by our measurements in that the difference between the two angles β_t and β_e was only 1.5 ± 0.83 degrees for jumps on wood. By contrast, the angle β_t was significantly higher than the angle β_e for jumps on the other two ground types (sand and soil). The differences between the three ground types stem from the specific conditions of the tibia on the ground (i.e., sliding or insertion). It was clearly illustrated in Figure 6 that the locust tibia firmly adhered to wood but often inserted into sand or slid on soil during the take-off process. The sliding or insertion altered the swinging angle of the hind leg, resulting in an altered take-off elevation angle. Although locusts can use their claws and tarsal adhesive pads to firmly stick to a few types of substrates [25,26], these strategies appear to be useless for successfully attaching to natural ground types, such as sand and soil. In addition, the significant difference in the angle γ resulted from the different β_t among the three types of ground following the quantitative relationship between these two angles established by Sutton and Burrows [6].

Another interesting finding is that only the aft component of the mass-specific reaction force (F_a) was significantly different among the three ground types while its total magnitude and other components were not (Figure 5). This finding might stem from the differing effect of the tibial angle on the three physiological components. Given that the total reaction force (F_t) is along the tibial axis, its aft and normal components were calculated by the decomposition theory of force as $F_t \cos(\gamma)$ and $F_t \sin(\gamma)$, respectively. Thus, a higher tibial angle leads to a reduced aft component and an elevated normal component. Although the F_t values did not significantly change among the three ground types, the aft component was increased by $0.35F_t$ (from $0.37F_t$ to $0.72F_t$) when the angle γ changed from 68 degrees (on wood) to 44 degrees (on sand) and by $0.26F_t$ (from $0.37F_t$ to $0.63F_t$) when the angle γ changed from 68 degrees (on wood) to 51 degrees (on soil). Similarly, the normal component decreased by $0.24F_t$ (from $0.93F_t$ to $0.69F_t$) when the angle γ changed from 68 degrees (on wood) to 44 degrees (on sand) and by $0.15F_t$ (from $0.93F_t$ to $0.78F_t$) when the angle γ changed from 68 degrees (on wood) to 51 degrees (on soil). Briefly, the influence of the angle γ on F_a was 1.5–1.7 times that of F_n, which might explain why significant differences only existed in F_a among the different ground types.

However, some limitations of this study require consideration. First, the sample size of jumps for each ground type was small, although paired statistical analyses were used in the comparisons. It should be emphasized that the results of this study are preliminary and should be used as a basis from which future studies examining the effect of ground type on locust jumps could be conducted. Second, the properties of the three ground types need to be quantitatively measured. Based on these measurements, the quantitative relationship between ground properties and locust jump behaviors could be characterized and provide some fundamental mechanical data for optimizing the jumping tactics of bioinspired robots on different ground types.

5. Conclusions

In this study, jumps of nine adult *L. m. manilensis* locusts on three different types of ground (sand, soil, and wood) were measured using a custom-developed test system. Specifically, measurements were made of the postural features, kinematics, and reaction force. Both the elevation angle β_t and tibial angle γ at take-off were significantly different among the three types of ground, which might have been caused by the hind legs slipping on or inserting into the ground. Nevertheless, the jumping kinematics (including the take-off speed and the mass-specific kinetic energy) were not significantly different among the different ground types, indicating that locusts were able to achieve robust jumping performance on the various substrates. This study provides preliminary data that contribute to enhancing our understanding of the jumping mechanisms in locusts, especially how jumping behaviors are adapted to different types of ground.

Supplementary Materials: The following are available online at http://www.mdpi.com/2075-4450/11/4/259/s1, Table S1: Experimental data of the kinematics for the locusts jumping on sand, Table S2: Experimental data of the kinematics for the locusts jumping on soil, Table S3. Experimental data of the kinematics for the locusts jumping on wood, Table S4: Experimental data of the reaction force for the locusts jumping on sand, Table S5: Experimental data of the reaction force for the locusts jumping on soil, Table S6: Experimental data of the reaction force for the locusts jumping on wood, Video S1: A typical video of the locust jumping on sand, Video S2: A typical video of the locust jumping on soil, Video S3: A typical video of the locust jumping on wood.

Author Contributions: Conceptualization, C.W. and Z.H.; experiment and formal analysis, R.C. and C.W.; writing—original draft preparation, C.W.; writing—review and editing, Z.H.; funding acquisition, C.W. and Z.H. All authors have read and agreed to the published version of the manuscript.

Funding: This work was supported by grants from the State Key Laboratory of Tribology (grant number: SKLT2015B06, SKLT2018B07) and the Beijing Institute of Technology Research Fund Program for Young Scholars (grant number: 3052019066).

Conflicts of Interest: The authors declare no conflict of interest.

References

1.   Scott, J. The locust jump: An integrated laboratory investigation. *Adv. Physiol. Educ.* **2005**, *29*, 21–26. [CrossRef] [PubMed]
2.   Heitler, W.J.; Burrows, M. Locust jump. 1. motor programme. *J. Exp. Biol.* **1977**, *66*, 203–219. [PubMed]
3.   Bennet-Clark, H.C. The energetics of the jump of the locust *Schistocerca gregaria*. *J. Exp. Biol.* **1975**, *63*, 53–83. [PubMed]
4.   Wan, C.; Hao, Z.; Feng, X.-Q. Structures, properties, and energy-storage mechanisms of the semi-lunar process cuticles in locusts. *Sci. Rep.-UK* **2016**, *6*, 35219. [CrossRef] [PubMed]
5.   Norman, A.P. Adaptive-changes in locust kicking and jumping behavior during development. *J. Exp. Biol.* **1995**, *198*, 1341–1350.
6.   Sutton, G.P.; Burrows, M. The mechanics of elevation control in locust jumping. *J. Comp. Physiol. A* **2008**, *194*, 557–563. [CrossRef]
7.   Santer, R.D.; Yamawaki, Y.; Rind, F.C.; Simmons, P.J. Motor activity and trajectory control during escape jumping in the locust *Locusta migratoria*. *J. Comp. Physiol. A* **2005**, *191*, 965–975. [CrossRef]
8.   Cofer, D.; Cymbalyuk, G.; Heitler, W.J.; Edwards, D.H. Control of tumbling during the locust jump. *J. Exp. Biol.* **2010**, *213*, 3378–3387. [CrossRef]
9.   Han, L.; Wang, Z.; Ji, A.; Dai, Z. The mechanics and trajectory control in locust jumping. *J. Bionic Eng.* **2013**, *10*, 194–200. [CrossRef]
10.   Chen, D.; Chen, K.; Zhang, Z.; Zhang, B. Mechanism of locust air posture adjustment. *J. Bionic Eng.* **2015**, *12*, 418–431. [CrossRef]

11. Gvirsman, O.; Kosa, G.; Ayali, A. Dynamics and stability of directional jumps in the desert locust. *PeerJ* **2016**, *4*, e2481. [CrossRef]

12. Gabriel, J.M. The development of the locust jumping mechanism. 1. allometric growth and its effect on jumping performance. *J. Exp. Biol.* **1985**, *118*, 313–326.

13. Gabriel, J.M. The development of the locust jumping mechanism. 2. energy-storage and muscle mechanics. *J. Exp. Biol.* **1985**, *118*, 327–340.

14. Duffy, B.M.; Harrison, J.F.; Kirkton, S.D. Growth within an instar reduces jumping performance in American locusts (*Schistocera americana*). *FASEB J.* **2008**, *22* (Suppl. 1). Available online: https://www.fasebj.org/doi/10.1096/fasebj.22.1_supplement.757.20 (accessed on 1 March 2008).

15. Arntzen, K.; Kirkton, S.D. Effects of oviposition on jump performance in the American locust (*Schistocerca americana*). *FASEB J.* **2008**, *22* (Suppl. 1). Available online: https://www.fasebj.org/doi/10.1096/fasebj.22.1_supplement.757.19 (accessed on 1 March 2008).

16. Carroll, J.P.; Kirkton, S.D. Effect of gravidity on jump performance and muscle physiology in the American locust. *FASEB J.* **2013**, *27* (Suppl. 1). Available online: https://www.fasebj.org/doi/10.1096/fasebj.27.1_supplement.714.4 (accessed on 1 April 2013).

17. Katz, S.L.; Gosline, J.M. Ontogenic scaling of jump performance in the African desert locust (*Schistocerca-gregaria*). *J. Exp. Biol.* **1993**, *177*, 81–111.

18. Hawlena, D.; Kress, H.; Dufresne, E.R.; Schmitz, O.J. Grasshoppers alter jumping biomechanics to enhance escape performance under chronic risk of spider predation. *Funct. Ecol.* **2011**, *25*, 279–288. [CrossRef]

19. Bennet-Clark, H.C.; Alder, G.M. The effect of air resistance on the jumping performance of insects. *J. Exp. Biol.* **1979**, *82*, 105–121.

20. Snelling, E.P.; Becker, C.L.; Seymour, R.S. The effects of temperature and body mass on jump performance of the locust *Locusta migratoria*. *PLoS ONE* **2013**, *8*, e72471. [CrossRef]

21. Wang, L.; Johannesson, C.M.; Zhou, Q. Effect of surface roughness on attachment ability of locust *Locusta migratoria manilensis*. *Wear* **2015**, *332–333*, 694–701. [CrossRef]

22. Mo, X.; Ge, W.; Romano, D.; Donati, E.; Benelli, G.; Dario, P.; Stefanini, C. Modelling jumping in *Locusta migratoria* and the influence of substrate roughness. *Entomol. Gen.* **2019**, *38*, 317–332. [CrossRef]

23. Sarmiento-Ponce, E.J.; Sutcliffe, M.P.F.; Hedwig, B. Substrate texture affects female cricket walking response to male calling song. *R. Soc. Open Sci.* **2018**, *5*, 172334. [CrossRef] [PubMed]

24. Burrows, M.; Morris, G. The kinematics and neural control of high-speed kicking movements in the locust. *J. Exp. Biol.* **2001**, *204*, 3471–3481. [PubMed]

25. Wang, L.; Zhou, Q.; Xu, S. Role of locust *Locusta migratoria manilensis* claws and pads in attaching to substrates. *Chin. Sci. Bull.* **2011**, *56*, 789–795. [CrossRef]

26. Woodward, M.A.; Sitti, M. Morphological intelligence counters foot slipping in the desert locust and dynamic robots. *Proc. Natl. Acad. Sci. USA* **2018**, *115*, E8358–E8367. [CrossRef]

MDPI
St. Alban-Anlage 66
4052 Basel
Switzerland
Tel. +41 61 683 77 34
Fax +41 61 302 89 18
www.mdpi.com

Insects Editorial Office
E-mail: insects@mdpi.com
www.mdpi.com/journal/insects